# HOW TO THINK LIKE
# STEPHEN
# HAWKING

By the same author:

*How to Think Like Sherlock*
*How to Think Like Steve Jobs*
*How to Think Like Mandela*
*How to Think Like Einstein*
*How to Think Like Churchill*
*How to Think Like Bill Gates*
*How to Think Like da Vinci*

# HOW TO THINK LIKE
# STEPHEN HAWKING

## DANIEL SMITH

Michael O'Mara Books Limited

**For Rosie and Lottie**

This paperback edition first published in 2020
First published in Great Britain in 2016 by
Michael O'Mara Books Limited
9 Lion Yard
Tremadoc Road
London SW4 7NQ

A CIP catalogue record for this book is available from the British Library.

Papers used by Michael O'Mara Books Limited are natural, recyclable
products made from wood grown in sustainable forests. The manufacturing
processes conform to the environmental regulations of the country of origin.

ISBN: 978-1-78929-225-1 in paperback print format
ISBN: 978-1-78243-561-7 in e-book format

1 3 5 7 9 10 8 6 4 2

Designed and typeset by Envy Design Ltd

Printed and bound by CPI Group (UK) Ltd, Croydon, CR0 4YY

www.mombooks.com

MIX
Paper from
responsible sources
FSC® C020471

# Contents

# Introduction

'His name will live in the annals of science; millions have had their cosmic horizons widened by his bestselling books; and even more, around the world, have been inspired by a unique example of achievement against all the odds – a manifestation of astonishing willpower and determination.'

MARTIN REES, *NEW STATESMAN*, 2015

When he was alive, Stephen Hawking was, quite simply, the most famous scientist on the planet. Only Einstein could claim to have achieved a similar level of fame in his own lifetime, but given the subsequent developments in mass communication, Hawking's celebrity arguably eclipsed even his. Diagnosed with the wasting disease amyotrophic lateral sclerosis (ALS, also commonly known as motor neurone disease) while still a student, Hawking spent several decades confined to a wheelchair. Towards the end of his life he struggled to control the muscle power required even to twitch his cheek. And yet, through it all, his mind blossomed and bloomed, casting light into the darkest corners of the universe. Since the publication of *A Brief History of Time* in 1988, he engaged with the wider public in a way no other scientist – alive or dead – ever came close to emulating.

He exponentially increased our understanding of how black holes work and, by so doing, has shed light on the very origins of our universe. Indeed, he brought us several steps closer to answering that ultimate question of science – where did we all come from? His investigations into the circumstances of the Big Bang – the cosmological event widely considered to have brought the universe into being – compelled us to re-evaluate everything we thought we knew. His radical theses not only challenged previous scientific orthodoxies, but challenged us to reconsider such fundamentals as the nature of time and the notion of a godhead.

There are those commentators who suggest that Hawking's fame clouded our view of the extent of his scientific achievements. True, Hawking was no 'Lone Ranger', but worked in collaboration with a multitude of very talented contemporaries during a golden age for cosmology. But whether he was the single leading light of his generation or merely registered among the 'elite of the elite' is almost beside the point. We might just as well ponder whether Alexander the Great outstrips Napoleon, or if Pelé bests Maradona. There will always be advocates to argue the case one way or the other, but we can never settle such arguments conclusively. Suffice to say, Hawking reshaped the scientific landscape and his influence will linger long into the future. It was, in all senses, a stellar career.

In the popular imagination, he will forever be the genius trapped in a failing shell of a body. It fulfils our desire for a romantic narrative to accompany the hard science. He was the man who was gifted with an extraordinary mind while his body betrayed him. In his overturning of vast medical odds – few medical professionals envisaged he would reach the age of thirty, let alone make it into his seventies – his story is at once tragic and heroic. His plight pulled at our heartstrings while filling us with hope that anything really is possible. In his own rather blunt words, 'No one can resist the idea of a crippled genius.'

Yet the reality is that Hawking did not neatly fit a storybook narrative. His intellect coexisted with a complicated personality that demanded us as a society to consider how we label 'the disabled'. Like his famous theoretical physicist forefather, Albert Einstein, he was not only a scientist but also a humanitarian. He campaigned on platforms as disparate, for example, as nuclear disarmament, disability rights and Arab–Israeli peace. Meanwhile – and again in common with Einstein – his private life was sometimes torrid. Two marriages and three children attest to an individual for whom love and relationships were a vital component of life. Yet his burning ambition and his commitment to his work – allied to a sometimes prickly disposition – left casualties along the way.

Even his ALS represented something of a paradox. While it vastly impaired the day-to-day quality of his life, it was his diagnosis while in his early twenties that arguably kick-started his career. Faced with premature death, he seized the opportunities that his talents opened up for him with a passion and drive that had not hitherto been apparent. In some lights, it can seem almost as if his disease played a part in liberating his intellectual imagination. One is put in mind of the words of Jean-Dominique Bauby, the author of *The Diving Bell and the Butterfly*, which documented his life after a stroke had rendered him the victim of locked-in syndrome (a condition he called his diving bell): 'My diving bell becomes less oppressive, and my mind takes flight like a butterfly.'

Strip away the science, which is admittedly sometimes forbidding, and Hawking emerged as a figure who allowed himself to dream of extraordinary things and who was blessed with a unique intellect that allowed him to make sense of those dreams. He could be brilliant and wise, just as he could be hot-headed and difficult. Underlying it all was an unstinting desire to make sense of the cosmos that we inhabit, a passion for knowledge that neither misfortune nor time (however you envisage it) can weary. Consider his almost *Peter Pan*-like words from 2013: 'So remember to look up at the stars and not down at your feet. Try to make sense of what you see and hold on to that childlike wonder about what makes the universe exist.'

It is apposite to say at this point that the object of this book is *not* to unpick Hawking's scientific theories. The man himself spent decades breaking down his work into digestible chunks for a wide readership. To believe that I could make a better fist of explaining, say, imaginary time than Hawking himself would be misplaced arrogance in the extreme. If not all of his popular writings are immediately 'gettable', that is less a reflection of his capabilities as a communicator than the sheer, brain-aching complexity of the material he was dealing with. In short, if you want a quick guide to his life's work, go to the source itself (starting with *A Brief History of Time*).

*How to Think Like Stephen Hawking* instead aims to get at the man behind the science and beyond the celebrity. While I will necessarily reference his science in the pages that follow, my aim is rather to explore his personality, how he went about forging his career, and what drove and inspired him. This is the eighth book in the *How to Think Like* series, and with each new title I never cease to be amazed at how the subjects exhibit not only startling abilities and character traits that set them apart, but also the frailties and character flaws that make them, ultimately, 'one of us'. I hope this volume will shed some light on the figure who came to epitomize 'science' for our age, but who also represented so many of the key facets of what it means to be human.

# Landmarks in a Remarkable Life

1942     Stephen William Hawking is born on 8 January in Oxford, England, to Frank and Isobel Hawking. The family return to their home in North London at the end of the Second World War.

1950     The Hawkings relocate to St Albans, about twenty miles north of London.

1953     Hawking wins a place at St Albans School. In his last year there he is accepted to study at the University of Oxford.

1959     He begins his undergraduate degree in natural sciences at Oxford's University College, his father's alma mater.

1962     He completes his degree, receiving a First. Hawking then moves to the University of Cambridge to commence postgraduate studies in cosmology. He also meets his future wife, Jane Wilde.

1963    He is diagnosed with amyotrophic lateral sclerosis (ALS).

1964    He attends a lecture by Birbeck mathematician Roger Penrose in London. Penrose's theories on singularities prove highly influential on Hawking.

1965    Hawking marries Jane Wilde.

1966    On finishing his doctorate, he is awarded a fellowship at Cambridge's Gonville and Caius College. Hawking undertakes research into singularities and black holes, working with Roger Penrose.

1967    Jane Hawking gives birth to a son, Robert.

1969    As his ALS worsens, Hawking begins to use a wheelchair.

1970    He reports that, in accordance with the rules of Einstein's general theory of relativity, the universe may have emerged from a singularity. Jane Hawking gives birth to a daughter, Lucy.

1971    Hawking shows how a black hole's event horizon expands with time. In the same year, he co-publishes the 'no-hair' theorem with Brandon Carter.

1973    He joins the staff at Cambridge's Department of Applied Mathematics and Theoretical Physics (DAMTP). Hawking and George

Ellis publish *The Large Scale Structure of Space-Time*. He is also co-author of four laws of black hole mechanics.

1974    He publishes a paper in the journal *Nature* entitled 'Black Hole Explosion'. It introduces to the wider world the theory that makes his name – Hawking radiation. He is also appointed Professor of Gravitational Physics at Cambridge.

1975    Hawking relocates his family to the USA so that he can work at the California Institute of Technology (Caltech). While there, he meets Kip Thorne, who becomes a long-time collaborator and friend. Meanwhile, Hawking is awarded the Pius XI Gold Medal for Science by the Vatican.

1977    Hawking and Gary Gibbons develop a revolutionary system of Euclidean quantum gravity.

1978    Hawking receives the Albert Einstein Medal.

1979    He is appointed to the prestigious chair of Lucasian Professor of Mathematics at Cambridge, a post previously held by Sir Isaac Newton, Charles Babbage and Paul Dirac. He is also made a fellow of the Royal Society. Jane Hawking gives birth to a second son, Timothy.

1982    He is made a Commander of the Order of the British Empire.

1983    Hawking publishes the 'no-boundary' theory in partnership with James Hartle, which describes how the universe could have emerged from nothing.

1985    A bout of pneumonia almost kills him while he is visiting the European Organization for Nuclear Research (CERN) in Switzerland. A resulting tracheostomy leaves him unable to speak naturally.

1987    He is awarded the Paul Dirac Medal.

1988    His popular science book, *A Brief History of Time: From the Big Bang to Black Holes*, is published and becomes an instant classic while breaking sales records.

1989    He is made a Companion of Honour.

1990    Hawking splits from his wife, Jane, and moves in with his former nurse, Elaine Mason.

1993    Publication of *Black Holes and Baby Universes and Other Essays*. Hawking also releases details of his research with John Stewart into 'thunderbolt singularities'. In addition, he becomes the first person to play himself in *Star Trek* when he appears in *The Next Generation*.

1995    Hawking and Elaine Mason wed.

1999    He joins with other leading public figures, including Archbishop Desmond Tutu, to promote a Charter for the Third Millennium, which demands improved rights for the disabled. He also cameos in *The Simpsons* for the first time.

2000    The police investigate allegations that Elaine Hawking has abused her husband. No charges are ever brought.

2001    Publication of *The Universe in a Nutshell*.

2002    Publication of *On the Shoulders of Giants: The Great Works of Physics and Astronomy*, as well as *The Theory of Everything: The Origin and Fate of the Universe*.

2003    Hawking is a vocal critic of the UK's part in the invasion of Iraq.

2004    He publicly concedes a bet made seven years earlier with John Preskill after accepting that information can escape from a black hole.

2005    Publication of *God Created the Integers* and *A Briefer History of Time*, the latter co-authored with Leonard Mlodinow.

2006    Hawking and Thomas Hertog propose a 'top-down' theory of cosmology. Hawking divorces Elaine.

2007    Hawking and his daughter Lucy publish the first of a series of children's fiction books about science. He establishes the Centre for Theoretical Cosmology at Cambridge and also undertakes a zero-gravity flight in the so-called 'Vomit Comet'.

2009    After thirty years he retires as Lucasian Professor of Mathematics. He is awarded the Presidential Medal of Freedom by Barack Obama.

2010    Publication of *The Grand Design* with Leonard Mlodinow.

2012    Hawking introduces the Paralympic opening ceremony in front of a worldwide television audience of millions.

2013    He receives the Special Fundamental Physics Prize, worth US$3 million. Publication of *My Brief History*.

2014    *The Theory of Everything*, a film based on Hawking's life, is nominated for the Academy Award for Best Film. It went on to win the award for Best Actor for Eddie Redmayne's portrayal of Hawking.

2015    Hawking assists in the launch of the Breakthrough Initiatives, a programme that searches for extraterrestrial life. In a lecture in Stockholm, Sweden, he also outlines a new theory on how information can escape a black hole.

2017    He is awarded an Honorary Doctorate from Imperial College London.

2018    On 14 March 2018 Hawking dies peacefully at home in Cambridge. In October 2018 his final book, *Brief Answers to the Big Questions*, is published.

# Follow Your Own Path

'If you understand how the universe
operates, you control it, in a way.'

STEPHEN HAWKING, *MY BRIEF HISTORY*, 2013

From an early age there were clear indications that Stephen Hawking had a natural inclination towards the sciences, although no one could have predicted that he would become the single most famous scientist alive on the planet. He was born on 8 January 1942 in Oxford, England, to where his parents, Frank and Isobel, had moved in the hope of escaping the German bombs then falling on London. His mother and father also had established ties with Oxford, both having studied at its famous university – Frank graduating in medicine and Isobel in philosophy, politics and economics.

Frank was a specialist in tropical diseases and had met his future wife while she was employed as a medical secretary in London. He spent some time in the pre-war period based in East Africa but returned to Britain around the outbreak of the Second World War. The authorities considered he would make a more significant contribution to the war effort by continuing his medical research work, rather than by entering the armed forces.

**DESTINED FOR GREATNESS?**

It has often been noted that Stephen was born on the day of the three-hundredth anniversary of the death of the brilliant Italian astronomer, Galileo. To some, this is a most portentous coincidence, although Hawking dismissed its significance out of hand, pointing out that some 200,000 other babies were also born that day, the vast majority of whom have lived their lives utterly unaffected by their 'connection' to Galileo. Significant or not, Hawking nevertheless followed in the Italian's footsteps more closely than any of those other babies.

Towards the end of the war, the family returned to the leafy North London suburb of Highgate, where they enjoyed a comfortable – although far from lavish – lifestyle. Their numbers were expanded by the arrival of a sister for Stephen, Mary, in 1943, followed by Philippa in 1947 and then an adopted brother, Edward, in 1954. By that time the Hawkings had relocated to St Albans, a historic town some twenty miles north of central London. Given their closeness in age, Stephen and his two sisters played together but also harboured a certain competitiveness. In fact, Stephen has described Mary as a brighter child than he was. For instance, while she had mastered reading by age four, he only learned when he was eight. He put his relatively late development in this area partly down to his parents' choice of school for him, the progressive Byron House School, which eschewed traditional methods of learning.

If he was not exactly bookish from the outset, he was nonetheless always interested in how things worked. He had a near obsession with model railways, for instance, and plundered his modest child's savings to buy an electric set (which performed, sad to relate, underwhelmingly). As he entered his teens, he displayed a similar love for model boats and aeroplanes, devoting a large part of his leisure time towards creating working models. Another of his hobbies was to invent board games, many of which were extraordinarily complicated. A war game, for example, required a board containing several thousand squares. In such ways, Hawking was creating mini universes that he was able to fully know and bring under his control. In other words, he was practising on a small scale for what would become his life's work.

The Hawking family provided an environment in which thinking differently was celebrated. Both parents were intellectuals who embraced a certain bohemianism. While that might not have seemed much out of place in 1950s Highgate, it certainly was in suburban St Albans (which Hawking described as having been a pretty staid place at the time). The family, for instance, enjoyed holidays in a gypsy caravan, which they kept on land near the coastal resort of Weymouth – at least, until the county council forced them to move it after several years. They also spent a year touring Europe and Asia in an old London black cab. On another occasion, Isobel (whose own freethinking had earlier seen her become a member of the Young Communist League) took the children to visit an old school friend in Spain who was married to the poet Robert Graves. Stephen subsequently shared a tutor with Graves's son for the summer.

In 1951 Hawking was an eager visitor to the Festival of Britain, a celebration of the modern world that the British government hoped might raise the nation out of its post-war doldrums. For the young model-mad lad from St Albans, it was nothing less than a revelation, turning him on to new forms of architecture, as well as the latest scientific and technological trends and developments. Over time, he established himself as a pivotal member in a circle of school friends who shared a passion for science. Tellingly, Stephen was nicknamed 'Einstein' by the gang. The group would discuss questions that ranged from how to build a radio-controlled aeroplane to how the universe began and whether there is a god or not. They might not have been the coolest kids in school – Stephen, skinny, unsporty and burdened with a lisp, was the very model of youthful nerdishness – but they were among the sharpest.

By the time it came to thinking about what he might wish to study at university, Hawking was adamant that his heart lay with physics (the most 'fundamental science') and astronomy, subjects he believed would help him 'fathom the depths of the universe'. This, though, was something of a disappointment to his father, who had hoped that Stephen would follow in his medical footsteps. But Stephen regarded physics, at least in some respects, as a more honest and meritocratic subject than medicine. For instance, many years later he would note that, 'I have never found that my lack of social graces has been a hindrance. But I think physics is a bit different from medicine. In physics, it doesn't matter what school you went to or to whom you are related. It matters what you do.'

The Hawkings had gone out of their way to promote independent thought, so Hawking Snr could hardly have been surprised when Stephen insisted on applying for natural sciences. Stephen's compromise came in his choice of institution – his father's alma mater, University College, Oxford. So it was that simultaneously Frank Hawking set his son on the path to scientific greatness, and Stephen asserted his will with the conviction that has characterized his life. Frank need not have worried anyway, since Mary would in due course become a doctor, so keeping that particular family tradition alive.

# Natural Talent Helps ...

'My practical abilities never matched up
to my theoretical enquiries.'

STEPHEN HAWKING, *MY BRIEF HISTORY*, 2013

That the Hawking family prided itself on its intellectualism was something noted by visitors to the household. Some, for instance, recall that mealtimes were often undertaken in complete silence, with each member of the family having their nose buried in a book catering to their particular interests.

Growing up in this climate, Stephen quickly became something of an autodidact. School provided him with a good, basic education, but he began to probe into disparate subject areas not covered by the conventional curriculum. That desire to understand 'how stuff works' soon had him moving on from taking apart gadgets and putting them back together again to ruminating on questions of scientific profundity that would perplex far older and more cultivated minds.

Not all of his investigations ended successfully. Indeed, if they had, we may assume that Stephen would quickly have tired of them. That he undertook projects a little beyond his reach is merely indicative that he was already prodding at the boundaries of what he might achieve. Furthermore, as the quote at the beginning of this section reveals, his real talent lay in performing somersaults of the mind, rather than perfecting a practical demonstration of them. If the young Hawking was intent on building a model aeroplane, you could be sure it would be underpinned by brilliant science, even if the finished product did not necessarily match up to that which existed in his mind's eye.

It was thus his inquisitiveness that truly marked him apart from his peers as a child. His yearning to be the first to solve problems that have stumped others has remained with him throughout his career. As he put it – somewhat indelicately – in a 2002 lecture at Cambridge's Centre for Mathematical Sciences, 'There's nothing like the eureka moment, of discovering something that no one knew before. I won't compare it to sex, but it lasts longer.'

He was also blessed with a naturally good memory, an attribute that he honed still further when the physical disabilities that befell him in adulthood rendered him ever more reliant on his mind working at optimum capacity. He was, for instance, renowned among his academic associates for his ability to memorize long equations (reputedly running to several pages) with unerring accuracy. While it is quite possible to train one's memory to become more efficient, it is difficult to envisage reaching a Hawking-esque level of recall without having an extraordinary base memory upon which to build.

Yet Hawking had little time for those who expect credit simply for the fact that they are born with an impressive intellect. Praise and reward should come only as a result of what one does with that intellect. To put it another way, there is little credit in rattling off a great long equation if you don't know how to manipulate that equation in order to open up a window on the world. Similarly, he had little time for the Mensa-style culture of celebrating so-called 'intelligence' for its own sake. Those who boast about their IQ (Intelligence Quotient – a score derived from standardized testing) he accused of being 'losers' – while conceding, it should be noted, that he hoped his own IQ was pretty high.

Nonetheless, there is little doubt that Hawking was born 'smart' – a fact that, as noted previously, prompted his schoolmates to nickname him 'Einstein'. Fortunately for the world, Hawking eventually came to realize that he was not content simply to be smart. Indeed, as we shall see, his ambitions extended even beyond being an Einstein Mk II. Like virtually every world-changer, he harboured an ambition to make a real dent in the universe and came to understand that he could not achieve such a grand aim through raw talent alone.

# … But It's Nothing Without Hard Work and Perseverance

'All my life I have lived with the threat of an early death, so I hate wasting time.'

STEPHEN HAWKING, SPEAKING AT THE 2013 PREMIERE OF THE DOCUMENTARY *HAWKING*

The journey Hawking took from prodigious natural intellect to the globally important theoretician he became was not always smooth. Until he began studying for his PhD, he was at serious risk of underachieving. His schoolboy nickname indicates that his potential was there for all to see, but the truth is that his academic achievement at school and then Oxford University fell short of what might be expected.

Hawking passed the Eleven-Plus exam – an examination then widely in use in the UK to determine what type of high school pupils should attend – and won a place at the respected St Alban's School, a grammar school catering to 'higher ability' students. He was immediately put into the top academic stream, but utterly failed to shine. At the end of his first year, his results placed him twenty-fourth in his class and he only narrowly avoided being moved down a stream. The following year was not much better, culminating in a twenty-third placing, followed by eighteenth the next year. Even as his results stabilized as he moved up the school, he was rarely placed more than halfway up the class. Furthermore, his strikingly untidy handwriting helped reinforce an impression among several of his teachers that the Hawking boy was a bit lazy.

They were almost certainly right – to a degree, at least. Such was the sophistication of his mind that he found much of what was being taught to him unchallenging and, dare it be said, rather dull. He achieved a respectable ten O level passes (O levels being the standard examination for sixteen-year-olds at the time) before studying for A levels (Advanced-level exams taken after a further two years of study) in maths, physics and chemistry.

He chose physics even though he considered the school syllabus too easy and obvious. It was, he realized, a necessary evil if he was to fulfil his dream of undertaking more advanced studies in astronomy and related subjects. These were the disciplines that gripped his imagination, since they would allow him to grapple with the big questions of the universe. Meanwhile, it was a maths teacher named Dikran Tahta who would emerge as the single most inspirational figure in Hawking's school life. Tahta, who would go on to lecture at Exeter University, brought a passion to his subject that fired Hawking's own enthusiasm. He fostered a climate of debate and urged his students to truly explore their intellectual horizons. In short, he wanted his pupils to unshackle their imaginations and not limit themselves by focusing only on passing exams.

---

**HAWKING THE HARDWARE ENGINEER**

Rather than mathematics and the sciences, Hawking's most marked successes at school resulted from his extra-curricular activities. Most impressive was his prominent role among a group of sixteen- to eighteen-year-olds who built a working computer (the Logical Uniselector Computing Engine or LUCE), utilizing equipment salvaged from an old telephone exchange. This at a time in the late 1950s when nobody other than a few heavyweight academic institutions and government departments possessed such machines – suggestive, perhaps, of the potential of the mind of our would-be engineer.

---

It is testament to Hawking's raw ability, if not his schoolboy commitment to studying, that he breezed through the Oxford University entrance exam, achieving a score of over 90 per cent in his paper. But even as an undergraduate at University College, his progress continued to be stifled by his belief that the course was too simple, which fostered his reluctance to knuckle down. By his own calculations, he averaged only about an hour of work a day during his time at Oxford, which is lightweight even for a budding genius.

Furthermore, Hawking was younger than many of his fellow students, a large number of whom had undertaken military service. As a result, he felt lonely and isolated for much of his time at Oxford, a situation he attempted to rectify by taking up rowing (he was the ideal size to act as a cox). As he strived to forge a social life, the imperative to throw himself into his work grew even weaker.

On one famous occasion, his tutor, Dr Robert Berman, assigned Hawking and three fellow students a series of thirteen fiendishly tricky problems. When the group reconvened a couple of weeks later, the others had struggled to answer more than one or two each. Hawking, by contrast, only began to study them on the day the answers were due but came back with nine of the problems solved. His instinctive feel for the subject was carrying him through and allowing him to get away with a minimal input of effort.

In 1962, as the end of his undergraduate course neared, Hawking was on the borderline between a First and an Upper Second Class (2:1) degree. Ultimately he was called in for a *viva* (an oral exam for those on the cusp between grades) and, according to his take on events, told his examiners that if they gave him a First he would get out of their hair and go to Cambridge to pursue a PhD. However, if they gave him a 2:1, they would be stuck with him in Oxford. Of course, he received his First. Berman suggested the reason was less to get Hawking off the Oxford scene than because the examiners 'had the intelligence to realize that they were talking to someone far cleverer than most of themselves'.

So it was on to Cambridge, with no indication that Hawking was any closer to finding his work ethic. Instead, there was every reason to think that his malaise was set to continue. As he would relate on the long-running BBC Radio 4 show, *Desert Island Discs*, in 1992, 'It was the end of the fifties, and most young people were disillusioned with what was called the Establishment ... I and most of my contemporaries were bored with life.'

Then something remarkable happened. Hawking got serious and got busy. The driving force for this reinvention, as we shall see in the next section and as is evident from the quote at the start of this one, was his declining health: Hawking thought that time was against him and was thus spurred into action. As we shall also look at in more detail, the arrival in his life of a young woman named Jane Wilde gave him further motivation to make the most of his talents.

Professionally, Hawking also managed to mine a seam of exceptional patience and (admittedly late-onset) self-discipline – a feat all the more impressive given the grave uncertainty about how long he could expect to live. By choosing to take on the most fundamental questions of our universe – problems whose answers had thus far evaded all probing – he embarked upon a career in which there is little in the way of shortcuts. The process of question-forming, devising hypothetical answers and then stress-testing them is epically time-consuming. All the more so given that cosmology does not lend itself to providing clear empirical evidence in the short-term. As he told the *Guardian* newspaper in 2005, 'It is no good getting furious if you get stuck. What I do is keep thinking about the problem, but work on something else. Sometimes it is years before I see the way forward. In the case of information loss and black holes, it was twenty-nine years.'

That we should make the most of what we have to give was a belief Hawking only truly began to foster in his twenties, but it came to be a guiding tenet of his adult life. In an interview in 2011, again with the *Guardian*, he was asked, 'So here we are. What should we do?' His response? 'We should seek the greatest value of our action.'

# Grasp the
# Big Picture

'I could see that the stars could draw him.'

AN OBSERVATION ON THE YOUNG HAWKING
BY HIS MOTHER, AS QUOTED ON *DESERT
ISLAND DISCS* IN 1992

Hawking's undergraduate degree at Oxford was in natural sciences, which in practice meant a lot of physics and maths. By his own admission, he was a rather more natural physicist than a mathematician. However, his true calling was cosmology, the subject that formed the basis of his postgraduate studies at Cambridge. But what does the subject of cosmology encompass and why was it so appealing to Hawking?

It may be argued that when he was starting out on his career, physics was split into two distinct camps. On the one hand were those determined to unravel the mysteries of the universe in its totality (the cosmologists), and on the other were those academics intent on looking at what happens at the sub-atomic level, in the burgeoning field of quantum mechanics. The 'big picture' guys and the 'detail' folk, if you like. Then there were those who strived to unite these often seemingly contrary disciplines, of whom Hawking was perhaps the leading light of our age (following in a tradition that started in earnest with – and ended in frustration for – Einstein). This is, of course, a grand oversimplification of the situation, but one with some merit for historians of twentieth-century science.

Such were his natural abilities that Hawking could have chosen to study particle physics for his PhD if he had so wanted. Instead he chose cosmology, a discipline that was just entering a golden age of theoretical development after a renewed interest. But its emergence was still so nascent that Oxford did not yet offer a course in it.

Cosmology, then, is the branch of scientific thought that seeks to understand the structure, laws and evolution of the universe, taking in everything from its fundamental elements to the nature of time and space. As Hawking recalled in *My Brief History*, when he began his studies at Cambridge, cosmology was a 'neglected field' that was 'ripe for development'. Back in the 1920s, Edwin Hubble had proved the galaxies are moving away from us, so providing evidence that the universe is expanding. Now the scientific community was intent on discovering how it all began and how it might end. After years of frustration at the lack of a challenge he found in his academic studies, Hawking now had a subject he could truly get his teeth into.

This was the kind of pure, unbounded science that ignited his imagination and he was determined to be at its forefront, discovering the laws that could explain the baffling wonders of our universe. A chance, we might say, for scientific alchemy. His enduring love for science as a tool to make the complex comprehensible was revealed in his 2011 interview with the *Guardian*. 'Science is beautiful when it makes simple explanations of phenomena or connections between different observations. Examples include the double helix in biology, and the fundamental equations of physics.'

# Stare into the Abyss

'Black holes ain't as black as they are painted.'

STEPHEN HAWKING IN A LECTURE AT THE
K.T.H. ROYAL INSTITUTE OF TECHNOLOGY
IN STOCKHOLM, 2015

At the root of Hawking's fame is his theoretical modelling of black holes – those areas of space where gravity is so strong that not even light is able to escape its grasp. Gravity is so strong in black holes because matter is condensed into a minute space – typically when a very large star collapses in upon itself as it dies. The prospects for anything that falls into a black hole are not good. Hawking's word for the process of extreme stretching and contorting that matter experiences inside a black hole was 'spaghettification'.

Scientists have postulated the existence of what we now call black holes since at least the eighteenth century, although it was not until 1915 that the German physicist, Karl Schwarzschild, demonstrated that Einstein's general theory of relativity provided the theoretical framework for their existence (a fact about which even Einstein himself was unconvinced). However, the two world wars curtailed further serious, co-ordinated research, but in the post-Second World War era, black holes were the subject of renewed interest. It is worth noting that the term 'black hole' was not coined until 1967, when Princeton alumnus John Archibald Wheeler introduced it; until then the less sexy phrase 'gravitationally completely collapsed stars' was in common use.

The study of black holes is particularly challenging since, as they do not emit any light, they are invisible. Effectively, we can only identify the presence of a black hole by noting its effects on surrounding stars. To study black holes, then, is the ultimate example of groping about in the dark. Hawking, metaphorically at least, shed more light on these astronomical curiosities than perhaps anyone else.

We now know that a black hole consists of an event horizon – that is to say, an outer surface inside of which nothing may resist the hole's gravitational drag – and a singularity, which is a point of infinite density that results from the implosion of an object when it collapses under its own gravitational pull. The traditional laws of nature suggest that a singularity is a physical impossibility, yet cosmology seems to demand their existence. Hawking would bridge this theoretical chasm and set the parameters for succeeding generations of researchers. His cosmic odyssey took a landmark turn in 1964 when he accompanied his PhD adviser, Dennis Sciama, to hear a talk by the renowned mathematician Roger Penrose. Penrose pondered the existence of a singularity at the heart of a black hole.

Hawking took Penrose's proposition on a step further by wondering whether singularity theory could be extended to the whole of the universe. By doing so, he virtually invented an entirely new – and extraordinarily rich – area of scientific research overnight.

## REWRITING THE BOOK

'A lot of prizes have been won for showing the universe is not as simple as we might have thought.'
STEPHEN HAWKING, *A BRIEF HISTORY OF TIME*, 1988

1970 would be Hawking's *annus mirabilis*, during which he firmly established himself on the scientific world stage. His big break came with the publication of a paper ('The Singularities of Gravitational Collapse and Cosmology') in collaboration with Roger Penrose, which showed how singularities inevitably result from the gravitational collapse inherent in Einstein's general theory.

Over the next few years he consolidated his growing fame with a series of theoretical breakthroughs. In 1971, for instance, he showed how the event horizon of a black hole increases with time. In the same year, he and Brandon Carter went public with their 'no-hair' theorem, which decreed that black holes are distinguishable from each other only by differences in their mass, charge and spin, and nothing else. In 1973 he was part of a team that demonstrated the correlation between the behaviour of black holes and the four basic laws of thermodynamics.

Then, in 1974, came arguably his single biggest contribution to scientific advancement. Hawking employed quantum theory to argue that black holes emit a steady stream of radiation (or, as he subsequently put it, '… black holes ain't so black'), with the rate of radiation increasing as the size of the hole decreases until the hole evaporates entirely. Crucially, then, it is possible for some matter to escape from a black hole, contradicting the established orthodoxy that nothing can break free of an event horizon.

Working in a newly emerging field, Hawking revelled in defining the rules as he went along. Always the free spirit, he took nothing as read and was prepared to challenge everything that the scientific establishment thought it already knew. As a result, scientists were compelled to start looking at black holes, and the universe itself, in an entirely new way.

# Tackle the Really Big Questions

'There is a singularity in our past which constitutes,
in some sense, a beginning to the universe.'

STEPHEN HAWKING AND G.F.R. ELLIS IN *THE
LARGE SCALE STRUCTURE OF SPACE-TIME*, 1973

Stemming from his attendance at Roger Penrose's lecture in 1964, Hawking's growing understanding of black holes informed his theoretical approach to surely the greatest question in cosmology – how did the universe begin? In attempting to answer that question, he elevated himself from star of the scientific firmament to global icon.

The original idea that the universe began with a 'Big Bang' (some 13.7 billion years ago, according to latest calculations) is widely credited to a Belgian priest, Georges Lemaître. There was a growing body of evidence in the late 1920s that our universe is actively expanding – not least from the British astronomer, Edwin Hubble, who, as previously noted, demonstrated that galaxies are speeding away from us in every direction. In 1927, Lemaître suggested that the expansion of the universe had an original starting point from a primordial atom.

Other scientists gradually built upon his work to develop a picture of the beginning of time. Adherents of the Big Bang theory envisaged a hugely dense, hot mass of energy and matter measuring no more than a few millimetres across. This existed, their thesis suggested, for just a fraction of a second at the beginning of the universe before this 'primeval atom' exploded, unleashing vast amounts of energy and undergoing a period of rapid expansion until conditions cooled and the universe as we know it today began its long evolution.

As Hawking started out on his academic life, the scientific world was split between those who were convinced that the Big Bang theory held the answers to our beginnings and those who preferred an alternative explanation. Chief among the Big Bang opponents was Fred Hoyle, who championed the steady state theory. This argued that, as the galaxies move away from each other, new matter is created so as to maintain a constant average density across the universe. According to this hypothesis, the universe thus stays basically the same (on the large scale) across all of time. Such a view necessarily rules out the possibility of a Big Bang moment.

Supporters of the Big Bang theory were bolstered in 1965 when cosmic microwave background radiation was recorded for the first time. Such radiation had long been predicted by modellers of the Big Bang, and its discovery seemed to bear out much of their work. Nonetheless, the tug of war between the Big-Bangers and steady-staters rattled on. The stage was thus set for Hawking and his revolutionary take on things.

**THE STUDENT BECOMES THE MASTER**

Ironically, Hawking had hoped that Fred Hoyle would be his PhD advisor when he applied to Cambridge. Instead, as time revealed, the student would come to undermine the master's teachings. In a further irony, it was Hoyle who is credited with introducing the phrase 'Big Bang' into public discourse – thus fixing in the public's mind an idea he was actually hoping to debunk.

From the seed of the 1964 Penrose lecture, Hawking conjured the idea that the universe might have emerged from a singularity. If a black hole could collapse in on itself to form a singularity, might not a singularity collapse outwards to create a universe? By 1970 he was convinced of the idea. In pursuing his thesis, he masterminded – or was at least pivotal in developing – theory after groundbreaking theory. He was, for instance, a notable champion of the now widely accepted 'inflation theory' (first laid out by Alan Guth in 1980), which argues that the universe underwent a rapid period of super-expansion in the nano-seconds after it was created. It is because of this super-expansion, many cosmologists believe, that our curved universe appears level to the naked eye (just as our own planet does to those of us upon it) and that physically distant areas of the sky look similar (since they were initially in close contact with each other).

Perhaps even more significantly, Hawking (with James Hartle of the University of California) published the 'no-boundary' theorem in 1983. This attempted to address how the universe could emerge from a singularity, from which – by definition – nothing can escape. Drawing on quantum theory to adjust the general theory of relativity, Hawking and Hartle redefined the terms of reference for space-time geometry in the early universe. Essentially, their equations did away with the need to consider what existed before the Big Bang. As Hawking argued, since the general theory proved space and time are not absolutes, but are dynamic quantities shaped by the universe's matter and energy, it makes no sense to talk of time before the universe began. Asking what came before the Big Bang, he said, is like 'asking for a point south of the South Pole'.

Hartle and Hawking's theory allowed cosmologists to predict the probability of the spontaneous creation of our universe, which then entered a phase of super-inflation. But it worked only in the context of what is known as a closed universe – which is to say, one with a positive curvature that expands like an inflating beach ball. However, as astronomical evidence grew in support of the idea that the universe may be flat or open, Hawking later reimagined the no-boundary theory with Cambridge colleague Neil Turok so that it could also be applied to these theoretical models.

Then, in 2006, Hawking introduced his vision of 'top-down' cosmology to the world – a thesis widely seen as an extension of the no-boundary theory. Developed with Thomas Hertog, with whom Hawking had worked at the European Organization for Nuclear Research (CERN), this approach suggests we are better placed to take the universe at its final (or rather, current) state and work backwards to model its past forms, rather than attempt to establish its initial state and then work forward. This opens the way for the 'many worlds' approach to cosmology (see page 58), with the job of the cosmologist becoming that of discerning which history best fits our astronomical data and from there calculating the probability of what the future of the universe will look like.

It is fair to say that Hawking's global fame ultimately lay with the verve and vision he brought to the question of how time began (or, perhaps we should say, didn't begin). But, in a broader sense, he was the world's best-known scientist because he embraced the biggest questions of them all, exploring fundamental existential conundrums with a fearlessness that sometimes came close to arrogance. He was a figure who constantly strove to 'fathom the far depths of the universe', as he phrased it in *Black Holes and Baby Universes and Other Essays*. But he put it simplest of all in his *My Brief History*: 'I am just a child who has never grown up. I keep asking "how" and "why" questions. Occasionally, I find answers.'

# Stand Upon the Shoulders of Giants

'It is … appropriate to acclaim Newton as the
greatest figure in mathematical physics.'

STEPHEN HAWKING AND WERNER ISRAEL IN
*300 YEARS OF GRAVITATION*, 1987

For all his public displays of self-deprecating humour, Hawking was not a man who underestimated his place in the scientific pantheon. It is notable that he did not receive a Nobel Prize, which many take to be a necessary accolade to be considered among the true elite (at least in the scientific realm). But more on the Nobel question later. Prize or no prize, commentators regularly place Hawking in a proud lineage of scientific greats and there is little to suggest that he himself did not feel he belongs there. The blurb for his own books, for instance, describes him as 'one of the most brilliant theoretical physicists since Einstein'.

And while Hawking acknowledged the inspiration that several of his famous forebears provided him with, he was also happy to critique their work. After all, his career was built on testing their theses and in many cases revising them so that they might fit with the latest astronomical data.

We can do worse than look to Hawking's own 2002 book, *On the Shoulders of Giants*, to gain an impression of who he regarded as his principal antecedents. Providing introductions to several of the key works of physics and astronomy from throughout history, he included the writings of Copernicus, Galileo, Kepler, Newton and Einstein.

Polish-born Nicolaus Copernicus (1473–1543) prompted a revolution in the way in which humanity regarded the universe (and crucially, our place within it) by creating a model that had the sun at its centre, rather than the Earth. It is little wonder that Hawking admired the way he overturned established dogma. Similarly, Galileo Galilei (1564–1642), the Italian who made such giant strides in our understanding of astronomy, mass and motion, was described by Hawking in a 2009 edition of *American Heritage of Invention & Technology* as 'perhaps more than any other single person ... responsible for the birth of modern science'.

Isaac Newton (1642–1727) – the English physicist and mathematician most famous for his mathematical elucidation of gravity, but who achieved much else of significance beside – is hailed in the quotation co-written by Hawking at the start of this section. In Hawking's view he was simply the crème de la crème. While Newton is supposed to have said, 'If I have seen further than other men, it is because I stood on the shoulders of giants,' Hawking credited him as the giant with the shoulders upon which all who followed have stood. And Einstein, Hawking said, 'is the only figure in the physical sciences with a stature that can be compared with Newton'. However, because Einstein's general theory was able to draw on the mathematical work of others (notably Bernhard Riemann), while Newton was compelled to 'develop his own mathematical machinery', Newton gains the edge in Hawking's opinion.

There are others who may be inserted into this chronology of the great physicists. We might, for instance, think to include the pioneers of quantum mechanics – figures such as Max Planck, whose groundbreaking work in the field earned him a Nobel Prize in 1918. He was among those theoreticians who wrestled with the mind-boggling notion that matter may exist as both a wave and a collection of particles contemporaneously. Such an idea rocked science to its foundations as uncertainty became the order of the day. One could only hope to calculate the probability, for example, of any particle being in a given place at a given time, but now you could not know for sure. Again, our world view was thrown into doubt.

In 1979, Hawking was appointed to the prestigious position of Lucasian Professor of Mathematics at Cambridge. It was a role previously filled not only by Newton but, in the early twentieth century, by another leading light in quantum physics, Paul Dirac. He too surely deserved a position in the grand lineage. Dirac, by coincidence, oversaw the PhD of Hawking's own

PhD supervisor, Dennis Sciama. But it was the former's work with Richard Feynman from the 1920s to 1940s – which birthed brilliant quantum field theories as they examined quantum mechanics and electromagnetic fields – that proved the most enduring influence on Hawking's own work. In an address Hawking made at the 1995 dedication of a memorial to Dirac in London's Westminster Abbey, he called him 'the greatest theoretical physicist since Newton'. After his own outstanding career, Hawking must have regarded himself as another candidate for that title.

## HAWKING AND EINSTEIN

'Einstein and Hawking earned their status as superstars, not only by their scientific discoveries but by their outstanding human qualities.'

FREEMAN DYSON, THE *NEW YORK REVIEW OF BOOKS*, 2011

Ever since his schoolmates presciently nicknamed him Einstein, it is to that icon of science that Hawking has been most regularly likened, although he shied away from the comparison. Some regarded him almost as a natural successor, but the picture is rather more muddied than that. Yet there are many striking parallels between the lives and work of the two men.

Hawking spent a significant portion of his career pursuing the so-called 'theory of everything' (see page 46), which seeks to establish a set of rules reconciling the general theory of relativity with quantum laws that often seem fundamentally out of kilter with one another. It was a goal that Einstein had also pursued and which ultimately defeated him. Hawking continued to seek it out until his death, while also acknowledging that there may not be an 'ultimate theory' anyway. Nonetheless, it is this fabled quest that has wedded together the careers of Einstein and Hawking in the popular imagination.

There were also echoes of Einstein in Hawking on a more personal level. Their natural inquisitiveness and extraordinary intellects failed to shine within the constraints of early, formal education. On a related note, both men found their mathematical knowledge sometimes lagged behind the advances in their theoretical thinking. In other words, both have at times struggled to conquer the maths that explains the phenomena they can visualize in their minds.

Just as Einstein excelled in the theoretical realm, so he was something of a liability in applied science, his unfinessed experiments on occasion proving a threat to life and limb in the lab. Hawking also always preferred to work within his imagination rather than in the laboratory, even before illness left him with little choice in the matter. It is no overstatement to say that they are among history's very finest exponents of the thought experiment (in other words, the examination of a thesis and its workability in the imagination rather than by practical testing).

Both, in addition, had broader concerns – from the question of whether there is a god to concerns over the proliferation of nuclear weapons, and a tendency towards socially liberal attitudes. Both also came to use their celebrity to garner support for causes they cared deeply for (see page 130). Then there is their shared experience of distinctly complicated personal lives, including divorces and remarriages, as if to prove that it is perhaps easier to navigate the mysteries of the universe than it is to master affairs of the heart.

Ultimately, though, they are united as enduring icons of science – an arena that traditionally throws up precious few figures who transcend academia to become recognized personalities in their own right. They each became the human face of scientific thought for their respective ages. You may not understand the general theory of relativity and you may only have got twenty pages through *A Brief History of Time*, but you could recognize both Einstein and Hawking in an identity parade and you know that, somehow, their work is both important and brilliant. That is, perhaps, all any scientist can ever hope for.

Writing in the *New York Review of Books* in 2011, esteemed theoretical physicist Freeman Dyson said of the duo: 'Both of them fit easily into the role of icon, responding to public adoration with modesty and good humour and with provocative statements calculated to command attention. Both of them devoted their lives to an uncompromising struggle to penetrate the deepest mysteries of nature, and both still had time left over to care about the practical worries of ordinary people. The public rightly judged them to be genuine heroes, friends of humanity as well as scientific wizards.'

# Be an Iconoclast

'So Einstein was wrong …'

STEPHEN HAWKING, DEBATING AT
CAMBRIDGE'S ISAAC NEWTON INSTITUTE FOR
MATHEMATICAL SCIENCES, 1994

While Hawking acknowledged the debt he owed to those who have preceded him, he was no respecter of reputation for its own sake. In Hawking's universe, nothing was sacred. His iconoclasm was a trait that showed itself sometimes in a grand gesture and at other times manifested far more subtly.

He had a reputation, for instance, of discreetly using his wheelchair against those who vexed him. He once ran over the toes of the Prince of Wales, although it is not clear whether Prince Charles had 'earned' the assault or was the victim of some errant vehicular control on Hawking's part. Nonetheless, when friends called out the scientist on the incident years later, he professed that he wished to be able to do the same to the then prime minister, Margaret Thatcher.

Working on the front line of academia for so long, Hawking ran up against plenty of professional colleagues with big reputations and rarely shied away from taking them on as he deemed necessary. Among his most notable 'victims' was Fred Hoyle, advocate of the steady state theory, against which Hawking, as the Big Bang disciple, argued fiercely. After overcoming his disappointment on learning that Hoyle would not be supervising his PhD when he moved from Oxford to Cambridge, Hawking soon came to be relieved that he did not have to work out of the Hoyle camp.

Shortly after arriving at Cambridge, Hawking befriended Jayant Narlikar, one of Hoyle's postgraduate research assistants. Over the course of their many conversations, Hawking began to suspect that Hoyle was erroneous in a particular respect of the research he was then undertaking. With a feel for the dramatic and an almost cavalier recklessness, Hawking decided to raise the issue with Hoyle at the end of a high-profile lecture the older man was giving at the famed Royal Academy in London. Hawking thus chose to bring down his (at the time) much more famous and senior colleague in a very public arena. If this was not an all-out intellectual assassination, it nonetheless brought to mind Brutus and Caesar on the Ides of March.

Hoyle, unsurprisingly, was not impressed by the episode, but he joined an impressive roll-call of famous names who fell victim to a withering Hawking attack. As the quotation on page 41 shows, not even Einstein entirely escaped his acerbic tendencies, on that occasion over the question of quantum uncertainty in the universe. Later on we will also look at the progressively entrenched position Hawking took on the question of whether there is a god at all, but suffice to say, spiritual icons were as ripe for critiquing as those from the secular realm.

Consider, for instance, his prickly response to receiving an award from the Pope in 1975 – the Pius XI Medal given by the Pontifical Academy of Sciences. Thanks to a rather simplistic reading of the Big Bang theory, there were those theologians who believed the thesis – in its championing of a 'moment of creation' – supported the notion of a creator god. As such, Hawking was not quite such an unlikely recipient of the award as we might assume today (although he, of course, never argued that his work supported such a position).

As chance would have it, he received news of the accolade just as he had been watching Jacob Bronowski's much heralded television series, *The Ascent of Man*. The particular episode he had just viewed included details of Galileo's trial and long years of house arrest, a punishment inflicted by a Vatican angered by Galileo's insistence that the Earth was not the focal point of the universe. Such was Hawking's retrospective ire at the papacy's behaviour that he was ready to turn down the award (for which Paul Dirac had recommended him), but he was subsequently persuaded to accept it. However, he was determined to make a stand on behalf of Galileo when he visited the Vatican to receive the medal. Risking a potentially uncomfortable scene, he demanded to view the account of his celebrated antecedent's trial, which was kept in the Vatican Library. His point had been made.

Nor did age temper his willingness to confront and undermine those targets he believed were deserving of such treatment. In 2011, speaking at Google's Zeitgeist Conference, he had the entire sphere of philosophy within his sights, claiming that the discipline was dead, not least because, in his opinion, philosophers have consistently failed to keep up with modern scientific developments. Scientists, he said, without so much as a hint of a blush, 'have become the bearers of the torch of discovery in our quest for knowledge'. Necessarily for someone who has spent his life questioning long-held beliefs and assumptions, Hawking had no time for sacred cows.

# Have No Limit to Your Ambition

'My goal is simple. It is a complete understanding of the universe, why it is as it is, and why it ends at all.'

STEPHEN HAWKING, AS QUOTED IN
*STEPHEN HAWKING'S UNIVERSE*,
JOHN BOSLOUGH, 1985

Among Hawking's most notable contributions to cosmology is his 'no-boundary' theory; indeed, his refusal to acknowledge boundaries of any kind is a trait we can trace throughout his career. This is no better illustrated than by his energetic pursuit of the so-called theory of everything, which aims to fill in the gaps that exist between Einstein's general theory and the world of quantum mechanics. It is a quest of Arthurian proportions that has stumped all who have come before, not least – as previously mentioned – Einstein himself.

As Hawking evolved his theory that the universe emerged from a singularity, he came up against the same problem again and again – the equations in Einstein's general theory break down when they are applied to a singularity. But because Einstein developed his theory to apply to the entire universe in all its vastness, it does not take account of the small-scale structure of matter that quantum mechanics deals with. So if you try to apply the general theory to the universe when it was but a tiny fraction of a millimetre across, the quantum structure must be taken into account. Hence the need for a theory of everything that weds the two disparate approaches together.

The very term 'theory of everything' is not without its problems. For starters, even if the general theory and quantum mechanics can be perfectly aligned, there are many who doubt that we will have 'all the answers' to the great mysteries. For instance, while it may comprehensively explain how we physically came to be here, will we be any closer to answering those great philosophical questions, like why are we here (if indeed there is a 'reason')? Nonetheless, it is the sort of phrase that immediately grabs the imagination (the 2014 Oscar-winning movie about Hawking's life took it for a title) and appeals to headline writers. Nor could Hawking the ringmaster be accused of having particularly discouraged its use as he strived to disseminate his work into the public sphere.

In 1980, for instance, in his inaugural lecture as Lucasian Professor of Mathematics, entitled 'Is the End in Sight for Theoretical Physics?', he claimed that scientists were in touching distance of the theory of everything. Indeed, he went so far as to predict that it would be discovered before the century was out. (Eighteen years later, in another public lecture, he acknowledged that 'it doesn't look as if we are going to quite make it'.) Even more famously, he concluded

*A Brief History of Time* with a spirited and memorable pitch for the theory:

> If we do discover a complete theory, it should in time be understandable in broad principle by everyone, not just a few scientists. Then we shall all, philosophers, scientists, and just ordinary people, be able to take part in the discussion of the question of why it is that we and the universe exist. If we find the answer to that, it would be the ultimate triumph of human reason – for then we would know the mind of God.

In pursuit of this Holy Grail of human knowledge, Hawking put a great deal of faith in string theory (or, to give it its full name, superstring theory) and especially in M-theory, which was born out of it. String theory, which established itself among a large corpus of theoretical physicists in the 1970s, envisages particles in a different way to the zero-dimensional point-like particles of traditional particle physics. In string theory, particles are modelled as one-dimensional vibrating strings, the vibrations defining their characteristics. The universe, according to string theory, is filled with these vibrating strings that constantly interact with each other. String theory gives rise to the notion of quantum particles carrying a gravitational force and provides a means of explaining, among other things, features of black holes that otherwise seemed to speak against the general theory.

M-theory emerged in the 1990s, principally through the work of Edward Witten. The 'M', he has suggested, might variously stand for 'magic', 'mysterious' or 'membrane'. Regardless of its etymology, it seeks to unify the five hitherto widely accepted versions of string theory into a single, overarching theory that envisages eleven dimensions. It is still incomplete and there is work to be done on creating a mathematical framework to fully explain it, but Hawking gave it a shot in the arm with the publication of *The Grand Design*, a book he wrote with Leonard Mlodinow in 2010 that outlines and expounds the theory.

Hawking was a wily operator and he knew that talk of the theory of everything was sure to gain popular attention. Yet he was also a realist. As the years went by, his public utterances suggested his faith that we will ever hit upon a complete, unified theory was diminishing. Take, for example, his preface to the 1993 collection of his selected works, *Black Holes and Baby Universes and Other Essays*. Already we see his bravura prediction of 1980 giving way to a far more cautious approach:

> The scientific articles in this volume were written in the belief that the universe is governed by an order that we can perceive partially now and that we may understand fully in the not-too-distant future. It

may be that this hope is just a mirage; there may be no ultimate theory, and even if there is, we may not find it. But it is surely better to strive for a complete understanding than to despair of the human mind.

In a 2002 lecture he gave entitled 'Gödel and the End of Physics', he went even further:

Some people will be very disappointed if there is not an ultimate theory that can be formulated as a finite number of principles. I used to belong to that camp, but I have changed my mind. I'm now glad that our search for understanding will never come to an end, and that we will always have the challenge of new discovery.

Justifying his strikingly new position, he cited the 'incompleteness theorems' of Austrian-American philosopher, mathematician and logician, Kurt Gödel.

Published in 1931, Gödel's highly complex thesis argued that any and all systems of logic (for instance, a given branch of mathematics) are necessarily incomplete, since there are always conceivable statements that may neither be proven nor disproven using the standard rules of that logic system. And should you seek to prove or disprove those statements by importing new rules or axioms into the system, ultimately you only create a new, larger system with its own set of unprovable statements. By following Gödel's logic, Hawking was forced to conclude that a true theory of everything can never be achieved. (Although, we might suppose Gödel's own logic must, according to his own argument, be incomplete, in which case his conclusions might yet be proven erroneous …) Regardless, Hawking referenced the incompleteness theorems not to deter those involved in the search for the theory of everything (of whom he remained one), but rather to suggest that perhaps we should not be too worried if that search does not culminate in knowing the mind of God after all.

# Imagine the Universe in Your Head

'I think in pictorial terms.'

STEPHEN HAWKING, *MY BRIEF HISTORY*, 2013

Hawking made the statement on the previous page in reference to *A Brief History of Time*. He said the book was in part an attempt to describe to a wider audience the mental images he carried round in his head using words and a few diagrams.

Having spent his adult life dealing with a diminishing ability to communicate verbally (whether orally or through the written word), it is perhaps little surprise that he processed ideas predominantly visually. Nonetheless, it is a conceptually difficult approach to grasp for those of us more used to fermenting our ideas in the crucible of traditional spoken or written language. While it seems eminently possible that Hawking's physical constraints led to an increased honing of his visualization skills (in the same way that there is evidence to suggest those who suffer from, say, blindness, develop a more acute sense of hearing), it also seems likely that Hawking was born with an innate ability to 'see' ideas and visualize concepts that very few of us can hope to emulate.

So it is, for instance, that Hawking was able to get to grips with the notion – as demanded by M-theory – that the universe comprises eleven dimensions. In a 2005 article in the *Guardian* newspaper (headlined 'Return of the Time Lord'), he acknowledged that this multi-dimensional take on the universe has its challenges: 'Evolution has ensured that our brains just aren't equipped to visualize eleven dimensions directly,' adding, 'However, from a purely mathematical point of view, it's just as easy to think in eleven dimensions as it is to think in three or four.' Likely the rest of us are left to ask 'Is it really?' as we struggle to get to grips with the standard three.

His extraordinary abilities to visualize were evident, for instance, in the work that put him at the forefront of the quest to understand quantum gravity in the 1970s. Along with Gary Gibbons, he developed a system of Euclidean quantum gravity that enables us to mathematically transform (using axioms of the ancient Greek mathematician, Euclid) the time dimension so that it correlates to the space dimensions. In this way, we can greatly simplify much of the mathematical work necessary to study, as an example, the characteristics of black holes. Euclidean quantum gravity was also vital to the formulation of the Hartle–Hawking no-boundary theorem.

Over his career, Hawking became the veritable poster boy for the concept known as 'imaginary time' – an idea that most of us can marvel at without truly comprehending. Hawking famously brought imaginary time to the masses in *A Brief History of Time*, but would later acknowledge it as 'the thing in the book with which people have most trouble'. He tried again to explain it in his 2001 book, *The Universe in a Nutshell*. First he depicted what we imagine as 'regular time' as a horizontal line running from the 'past' at one end to the 'future' at the other. 'Imaginary time', by contrast, runs perpendicular to 'regular time' so as to take on the aspect of other spatial dimensions. It is not, then, imaginary in the sense that it is not real or that it has been made-up. 'From the viewpoint of positivist philosophy,' he explained, 'one cannot determine what is real. All one can do is find which mathematical models describe the universe we live in. It turns out that a mathematical model involving imaginary time predicts not only effects we have already observed but also effects we have not been able to measure, yet nevertheless believe in for other reasons. So what is real and what is imaginary? Is the distinction just in our minds?'

In other words, for Hawking, imaginary time is simply another tool to ease the job of making sense of the reality around us. A way, as he once described it, to sidestep the thorny scientific and philosophical notion of time having a beginning by turning it into a direction in space. However, he subsequently reassured those still struggling with the precept that it is not truly necessary to understand imaginary time in all its detail in order to grasp its significance to his work, but just to realize that it is different from what we call real time. Thankfully, he could see it all mapped out in his mind even when we couldn't.

This extraordinary 'third eye' brings to mind more connections between Hawking and Einstein. The latter, for instance, once noted, 'I very rarely think in words at all. A thought comes, and I may try to express it in words afterwards.' It might have been Hawking talking about *A Brief History of Time*. On another occasion, writing in 1945, Einstein stated that 'the words or the language, as they are written or spoken, do not seem to play any role in my mechanism of thought'. Instead he described 'psychical entities which seem to serve as elements in thought' that were 'more or less clear images which can be "voluntarily" reproduced and combined'.

The nature of their respective work meant that both Einstein and Hawking laboured – sometimes literally – in the dark. Such were the depths of the mysteries that they took on that only the most truly inventive and visionary minds stand any chance of success. Both could claim to be almost super-humanly blessed in that department. As Einstein commented in 1929: 'Imagination is more important than knowledge. Knowledge is limited. Imagination encircles the world.'

## THE MANY-WORLDS INTERPRETATION

'There might be one history in which the moon is made of Roquefort cheese. But we have observed that the moon is not made of cheese, which is bad news for mice.'

STEPHEN HAWKING, *THE GRAND DESIGN*, 2010

In a consideration of Hawking's mental gymnastics, it would be remiss to ignore the way he built upon the so-called many-worlds interpretation (MWI) – and Richard Feynman's 'sum-over-histories' in particular.

The MWI had life breathed into it by American physicist Hugh Everett III in the late 1950s and was given its distinctive name by fellow American academic, Bryce Seligman DeWitt, a few years later. In (very) brief, the hypothesis argues that there is a potentially infinite number of universes, so that everything that might have happened in the past in our own universe has happened in another parallel one, giving rise to the existence of any and all histories and futures. Reality, in other words, becomes multi-branched, with our reality equating only to the particular branch we exist within. In terms of quantum mechanics, this allows for every potential outcome to be realized, thus removing 'randomness' from the quantum world.

Hawking utilized the MWI as a tool to calculate the probabilities of a hypothetical Condition A existing given the observable existence of Condition B. Whether all those parallel universes exist as a physical reality becomes, to an extent, neither here nor there, since their theoretical existence alone served his purposes. In terms of his quest to determine how the universe began, Hawking made special use of the work of Richard Feynman (a man Hawking described as a 'colourful character', not least because Feynman had a penchant for playing the bongos at a strip bar in Pasadena when not being a brilliant physicist) and his 'sum-over-histories' (also known as 'path-integral formation of quantum field theory'). Hawking explained it thus in the J. Robert Oppenheimer lecture he gave at the University of California, Berkeley, in 2007:

In order to understand the origin of the universe, we need to combine the general theory of relativity with quantum theory. The best way of doing so seems to be to use Feynman's idea of a sum over histories … He proposed that a system got from a state A, to a state B, by every possible path or history. Each path or history has a certain amplitude or intensity, and the probability of the system going from A to B is given by adding up the amplitudes for each path. There will be a history in which the moon is made of blue cheese, but the amplitude is low, which is bad news for mice.

By applying Feynman's sum-over-histories to Einstein's take on gravity, Hawking explained in *A Brief History of Time* that 'the analogue of the history of a particle is now a complete curved space-time that represents the history of the whole universe'. In other words, the bridge between the general theory and quantum mechanics edged ever nearer thanks in no small part to Hawking's ability to capture the universe in his mind's eye.

# Don't Let Misfortune Define You

'I think people in the disabled community would
say he's the biggest celebrity going. And he's
shown that having a disability is no obstacle
to achieving great things.'

DR TOM SHAKESPEARE, DISABILITY RIGHTS
CAMPAIGNER, 2015

Hawking was twenty years old with the world at his feet when the first signs of serious illness began to manifest themselves. The first thing he noticed was that he was no longer able to row a sculling boat properly and he also began to struggle to tie up his shoelaces. When he returned home to St Albans for Christmas in 1962, he went ice skating, fell and was barely able to get back to his feet – the first time his mother realized that something was amiss. Friends present at his twenty-first birthday party in January 1963 observed how he was unable to pour drinks for them properly. After a tumble down some stairs, he visited a doctor, who told him to ease his alcoholic consumption. But his mother was by then convinced there was more to it than that and instigated further testing. Eventually he was diagnosed with amyotrophic lateral sclerosis (ALS) – a degenerative wasting disease. The prognosis was about as grim as it could be. He could expect to be dead before his mid-twenties.

Hawking was, naturally, left reeling and entered a

dark psychological period as he was forced to confront his own mortality. Although he denied later reports that he began to drink heavily, he found solace in the music of Wagner and became what he has described as 'somewhat of a tragic figure'. Yet he emerged from it more focused and more determined than ever to make his mark on the world. Any indignation he might have felt at being struck by such an unforgiving condition was set aside in pursuit of achievement. And when his demons did appear, he thought back to a boy who had occupied a hospital bed opposite him before he succumbed to leukaemia. Whenever he was feeling self-pity, Hawking would tell himself that there is always someone else worse off.

Pre-diagnosis, Hawking's studies had been drifting somewhat. Now he embarked on his research with renewed vigour. He had a 'road to Damascus' moment, realizing that there were lots of worthwhile things he could do if he were reprieved and that if he was going to die anyway, he 'might as well do some good'. In the words of Nathan Myhrvold (who worked with Hawking before making his fortune by selling his computer business to Bill Gates and becoming a key figure in the rise of Microsoft), 'Stephen once tried to convince me his illness was an advantage because it helped him to concentrate on the important things.'

To his surprise, Hawking found that in many respects he was enjoying post-diagnosis life more than the pre-diagnosis version. This was in no small part down to the arrival in his life of Jane Wilde, the woman who would become his first wife. Buoyed by her support, he also began to make serious headway in his scientific investigations. So began the ascent of his remarkable career.

There were certain practical advantages, too. Hawking long acknowledged that his career benefitted because he had not had to undertake the arduous volume of teaching and committee-sitting that many of his able-bodied colleagues faced. His disability also left him with large amounts of time in which he was able to do little else but think. As Professor Brian Cox would tell journalists in 2012: 'One thing he told me was that he spends a great amount of time theorizing as he gets in and out of the bath, simply because it takes so long.' It is indicative of the way that Hawking could take what might be a source of immense frustration and transform it into something constructive and valuable.

Against the backdrop of deteriorating health over more than five decades, Hawking's star continued to rise. From promising postgrad student at Cambridge, he became a world-renowned theoretician, a bona fide superstar of the science firmament, a global celebrity and an icon of the age. It was, by any standards, an extraordinary achievement, regardless of the additional challenges of his physical incapacitation.

But when questioned as to how he coped with his disability, he was always quick to emphasize the upsides of his life. His quality of life was good, he said, despite his condition, and he did most of what he wished to do. Certainly, it is difficult to imagine how he could have achieved much more even if his health had been impeccable. His one regret, he conceded, was a deeply personal one – that he was not able to play with his children and grandchildren as he would have wished.

### A MANIFESTO FOR THE ABLE IN SPIRIT

In a 2011 interview with Claudia Dreifus for *The New York Times*, Hawking laid out his manifesto for those suffering from serious physical impairment: 'My advice to other disabled people would be, concentrate on doing things your disability doesn't prevent you doing well, and don't regret the things it interferes with. Don't be disabled in spirit as well as physically.' In many ways, the world defined Hawking at least partially in terms of his disability, but Hawking himself refused to let it define him.

Hawking was forced to play to his strengths, but did so with unstinting effectiveness. His single-minded attention to the areas in which he was able to excel is hinted at in an article he wrote in 1984 for *Science Digest* entitled 'Handicapped People and Science'. 'I am afraid that Olympic Games for the disabled do not appeal to me, but it is easy for me to say that because I never liked athletics anyway. On the other hand, science is a very good area for disabled people because it goes on mainly in the mind.'

By 2012, though, it was apparent that he had softened towards the athletic spectacle when he kicked off London's Paralympic opening ceremony with these inspiring words: 'The Paralympic Games is about transforming our perception of the world. We are all different, there is no such thing as a standard or run-of-the-mill human being, but we share the same human spirit. What is important is that we have the ability to create. This creativity can take many forms, from physical achievement to theoretical physics. However difficult life may seem, there is always something you can do and succeed at.'

Hawking embodied the ultimate story of triumph over adversity. While his body failed him in the most fundamental ways, he did not allow it to hold back his extraordinary intellect. He was proof that the human body is merely a shell, unable to shackle the flights of the mind or an indomitable spirit. In his *Desert Island Discs* interview in 1992, he summed up his situation in this understated manner: 'I don't regard myself as cut off from normal life, and I don't think people around me would say I was. I don't feel a disabled person –

just someone with certain malfunctions of my motor neurones, rather as if I were colour blind.'

## WHAT IS AMYOTROPHIC LATERAL SCLEROSIS?

'Mere survival would have been a medical marvel, but of course he hasn't merely survived.'

MARTIN REES, *NEW STATESMAN*, 2015

Amyotrophic lateral sclerosis is known in the USA as Lou Gehrig's Disease, in recognition of the famed New York Yankees baseman who was diagnosed with and killed by the illness before his thirty-eighth birthday. ALS is also sometimes referred to as motor neurone disease, although this is an umbrella term for a group of progressive neurological disorders of which ALS is just one.

The disease attacks the motor neurones (nerve cells) present in the brain, brain stem and spinal column that control voluntary muscle movement. As the disease progresses, so the sufferer becomes incapable of controlling muscles, which in turn waste away. ALS is an incurable affliction, with death commonly resulting from a patient's inability to swallow or breathe independently. While ALS attacks cells in the brain, it does not impair intellect. There is no clear pattern of associated risk factors among those who contract ALS, although some 5 to 10 per cent of cases are hereditary.

The initial prognosis that Hawking would likely be dead within two years of his diagnosis was not unduly conservative. The average lifespan for sufferers is between two and five years, with about a fifth of patients making it past five years. Only about 5 per cent make it beyond twenty years, which gives an indication of the extraordinary longevity that Hawking achieved. When he died in March 2018, he was believed to have been the longest-lived sufferer known to medical science. As to how he managed to live so long with ALS is a mystery, although some experts believe that those who are struck by the disease early (the average age at diagnosis is fifty-five or so) somehow adapt better.

# Overturn the Odds

'My expectations were reduced to zero when I was twenty-one. Everything since then has been a bonus.'

STEPHEN HAWKING, QUOTED IN *THE NEW YORK TIMES*, 2004

Hawking's life has now been depicted on screen several times. His is a story that, were it not true, might be accused of pushing the realms of credibility. A talented if rather lazy student is diagnosed with a terminal illness that gradually encases his mind in a prison of his own body. However, mining hitherto unseen reserves of determination, he establishes himself as the most celebrated scientist on the planet. Which is to say nothing of his private life that has, for better and worse, rarely been short on dramatic interest.

It makes for a great elevator pitch – a sort of geek's version of *Rocky*, with less boxing and more theoretical physics. At its core is the hero who rises to the most intimidating challenges that life can throw at him. Yes, Hawking's fame rests on his brilliant intellect, but it is difficult to imagine that he could have explored and exploited that intellect in the way that he was able to, given the problems he faced, were it not for his remarkable fortitude and courageousness.

It has been pointed out, with some justification,

that sometimes his determination could manifest itself in not entirely appealing ways. He could be, it has been said, short-tempered, forthright to the point of aggression, and arrogant towards those colleagues he regarded as intellectually inferior. He was undoubtedly personally ambitious, rarely deigning to undersell himself and eager to attract the economic rewards and public accolades that he believed his work justified. Yet how far might his star have risen, we might ponder, if he had lacked his steely core, given the hand that life dealt him just as his career was beginning? Not very far, we might conclude.

When he was diagnosed with ALS aged twenty-one, Hawking freely admitted he felt self-pity. Any other response would seem almost unnatural. It was, he considered, entirely unfair and he felt cheated that it should have happened to him. But after that initial slump in his spirits, Hawking embraced work and life with a renewed energy. With each progressive medical setback, he found a way to work around it so that he could create the space for his intellect to weave its magic.

As an individual who suffered symptoms that many people would consider unbearable, he was regularly invited to join in the debate on assisted dying. In broad terms, he defended the right of any individual to end their life if they so desired. However, as the People's Daily Online reported in 2006, his position was rather more nuanced. While supporting the concept of voluntary euthanasia, he also considered that to take that option was 'a great mistake'. 'However bad life may seem,' he was quoted as saying, 'there is always something you can do, and succeed at. While there's life, there is hope.'

He had perhaps greater moral authority than almost anyone else to make such a statement. Just as the symptoms of his ALS became more acute with the passing years, so he continued to confound anyone who dared write him off. When on a trip to the USA in the mid-1960s, his speech rapidly deteriorated so that only those who knew him well could decipher his slur. In order to communicate, he used translators to assist him. By the end of the decade he was confined to a wheelchair, with Jane having to assist him with day-to-day functions. As his situation got worse in the 1970s, the couple brought in additional live-in help so that he could continue to work and operate in public life.

Come the 1980s, he almost died while visiting CERN in Switzerland. With her husband suffering from a blocked windpipe and pneumonia, Jane agreed that he be given a life-saving tracheostomy (in which a breathing device is implanted into a hole cut in the neck) that removed the last remnants of his natural voice. Refusing to buckle under the strain or retreat from the life he had built for himself, he instead co-operated in the development of revolutionary new technology.

Voice software and infrared remote controls gave him the ability, for example, to control electrical equipment, open and shut doors and vocalize his thoughts merely by twitching his cheek. While others might have concluded that the game was up, Hawking used his condition to inspire scientific advancement. (Although he refused to 'upgrade' his simulated voice, recognizing that the robotic, Americanized accent he was initially given has become something of a trademark.) There were several other close calls with death too, including a sharp deterioration in his health in 2009, but time and again he astonished his doctors by bouncing back from the brink.

Even after his death, he remains an icon of what might be achieved by those with disabilities, a symbol of hope that ability can trump disability. And he knew that determination was a key ingredient in his achievements. Interviewed on BBC Radio 4 in 1992, he observed, 'Anyone with any nous is called stubborn at times. I would prefer to say I'm determined. If I hadn't been fairly determined, I wouldn't be here now.'

## KEEPING A SENSE OF HUMOUR

'Life would be tragic if it weren't funny.'

STEPHEN HAWKING, QUOTED IN *THE NEW YORK TIMES*, 2004

With the best will in the world, physics has not historically been much associated with humour and wit. Yet it was common for those who met Hawking – often expecting to encounter some ethereal presence – to be struck by his fulsome sense of humour, which could range from self-effacement to withering sarcasm. In a 2014 interview with *Vanity Fair* magazine, Eddie Redmayne (who famously received an Oscar for depicting Hawking in *The Theory of Everything* but whom, Hawking teased, did not boast his good looks) commented:

> The overriding thing when I met [Hawking] was this extraordinary, razor-sharp wit and formidable humour. There is a kind of mischief, a Lord of Misrule quality that I tried to take away from spending time with him, that I tried to bring into the performance.

Belying the image of the serious über-intellectual, Hawking long realized the importance of seeing the funny side of life. Indeed, he regarded it as a vital contributory factor to his unexpected longevity, as when he told a PBS television special in 2014: 'Keeping an active mind has been vital to my survival, as has been maintaining a sense of humour.' Of course, Hawking faced particular practical challenges when trying to crack a joke – namely, that it was not within his powers to drop an off-the-cuff one-liner into conversation, since the process of constructing a sentence to put through his voice synthesizer took minutes at least. However, when he did unleash his mischievous sense of humour, it was all the more disarming.

For example, one of his former nurses remembers the unexpected twist her interview took when she applied to work for him. A recent graduate from nursing school, the young woman was going for her first job and, unsurprisingly, felt in awe of the man she was hoping to care for. She stayed up into the small hours preparing herself for their meeting and set off for the appointment feeling about as prepared as she could, in expectation of a thorough interrogation. Instead, he asked her a single question: 'Can you poach an egg?' When she replied in the affirmative, he instantly gave her the job. The incongruous exchange took her off guard but immediately broke the ice between them.

Nor was he afraid to make a joke at his own expense. On several occasions, he told strangers who recognized him when he was out and about that he was not the real Stephen Hawking who is, he added, much better looking. At other times, though, he used his wit to go on the offensive. For instance, in his lecture 'Is Everything Determined' (delivered at a Sigma Club seminar in Cambridge in 1990), he argued against those who claim that all existence is predetermined with a funny and incisive observation: 'I have noticed that even people who claim everything is predetermined and that we can do nothing to change it look before they cross the road.'

Then there was that side of him that enjoyed simply goofing about. We saw him lend his comedic talents to such shows as *The Big Bang Theory*, *Futurama* and *The Simpsons*. And when, while being taken for a tour round the set of *Star Trek* in 1993, he saw the USS *Enterprise*'s legendary 'warp drive' (a hypothetical faster-than-light propulsion system), he jested, 'I'm working on that.'

Humour also became a tool-in-trade that allowed him to break down barriers when he was speaking on the public stage. In 2015, for instance, he was delivering a talk at the Sydney Opera House, shortly after the world's biggest boy band of the moment had announced a planned hiatus. A member of the audience posed the question: 'What do you think is the cosmological effect of Zayn [Malik] leaving One Direction and consequently breaking the hearts of millions of teenage girls across the world?' Hawking deadpanned, 'Finally, a question about something important.' Then he gave his considered response:

My advice to any heartbroken young girl is to pay close attention to the study of theoretical physics. Because one day there may well be proof of multiple universes. It would not be beyond the realms of possibility that somewhere outside of our own universe lies another, different universe, and in that universe, Zayn is still in One Direction.

It says much of society's preconceptions that Hawking's wit often came as a surprise. 'Genius', disability and levity are often regarded as unlikely companions. But as his daughter Lucy put it to *Grazia* magazine in 2015:

People sometimes forget that people with disabilities have personalities. I like it when Dad's sense of humour comes across. When you think of what he goes through in terms of day-to-day survival, his sense of humour probably helps keep him alive.

# Listen Like Hawking

'Music is very important to me.'

STEPHEN HAWKING, *DESERT ISLAND DISCS*, 1992

In the face of the physical constraints under which he lived, Hawking managed to find great solace in music. Indeed, in the interview from which the above quotation comes, he described physics and music as his 'two main pleasures'. For starters, he could explore his fascination with both within his mind, so that his physical disabilities became an irrelevance. But while physics set his mind racing, music served as a release – and could be enjoyed in a far less pressurized context.

He was in particular a fan of opera and classical music, although he enjoyed the occasional slice of mainstream pop, too. In an 'Ask Me Anything' interview on Reddit in 2015, he cited his favourite song ever as Rod Stewart's 'Have I Told You Lately' – a guilty pleasure he was surely not alone in sharing. And, as a youth, he was an avid listener to the Top Twenty chart as revealed by Radio Luxembourg on Sunday evenings, being a particularly big fan of The Beatles, who 'came as a welcome breath of fresh air to a rather stale and sickly pop scene'.

However, in general he was likely to be found listening

to something a little more highbrow. For his fiftieth birthday, for example, he was given the complete works of Mozart on CD – amounting to more than two hundred hours' worth of listening. When he appeared on *Desert Island Discs* – a show that requires its guests to create a list of the eight pieces of music they would wish to have with them in the event that they were stranded on a desert island – Hawking used up one of his selections on Mozart's hauntingly beautiful *Requiem*, which remained unfinished at the composer's death. Also included was the String Quartet No. 15 in A minor, Op. 132, by Beethoven (1770–1827). Just as Hawking's star shone despite the onset of disability in adulthood, so Beethoven produced some of his finest work having been afflicted with deafness, a condition that we might have expected to mark the end of his serious labours. The other classical works on Hawking's list were the Violin Concerto in D Major by Johannes Brahms (1833–97) and the *Gloria* by Francis Poulenc (1899–1963).

In terms of operatic composers, Hawking was a confirmed fan of, among others, Christoph Gluck (1714–87) – whose most celebrated work is *Orfeo ed Euridice*; Giuseppe Verdi (1813–1901) – whose operas include such staples of the genre as *Aida*, *Rigoletto*, *Il trovatore* and *La traviata*; and Giacomo Puccini (1858–1924). In fact, the latter found his way on to the desert-island list with 'O Principe, che a lunghe carovane' from his last opera, *Turandot*.

But among all the great operatic composers, he was perhaps most profoundly attached to Richard Wagner (1813–83), whose immersive and richly textured pieces – as has already been mentioned – became a companion to Hawking in the weeks and months that followed his ALS diagnosis. Wagner was, according to Hawking, 'someone who suited the dark and apocalyptic mood I was in'.

---

**HAWKING AND WAGNER**

Coincidentally, Wagner would have a role in another period of medical uncertainty for the physicist. In 1985 Hawking was preparing to travel to see the composer's magnum opus, the *Ring of the Nibelung* cycle, at Wagner's spiritual home in Germany – Bayreuth – when he was struck down with an illness that almost killed him and resulted in his tracheostomy. Those events, however, did little to diminish his respect for Wagner, whose *Die Walküre* (*The Valkyrie*, debuted in 1870) – the second of the *Ring Cycle* operas – was also featured among his desert-island discs.

---

As for operatic performers, Hawking was known to be an admirer of Angela Gheorghiu, the Romanian soprano who took her professional bow in 1990 and has gone on to wow audiences from New York's Metropolitan Opera and London's Royal Opera House to Milan's La Scala. It is also worth noting that Hawking included French chanteuse Édith Piaf (1915–63) and her most famous song, 1959's 'Non, je ne regrette rien' ('No, I regret nothing') among the tracks he would take to his desert island. No doubt the defiant lyrics so evocatively performed by the tragedy-stricken 'Little Sparrow' touch a chord with the physicist who faced more than his own fair share of tribulation.

# Two Heads Are Better Than One

'Speech has allowed the communication of ideas,
enabling human beings to work together to
build the impossible.'

STEPHEN HAWKING, BRITISH TELECOM
ADVERTISEMENT, 1993

While Hawking serves as a beacon of individual brilliance, he generally worked in a collegiate atmosphere and forged close alliances with fellow academics. Much of his most important work, as will become evident, arose from formal collaborations.

It is a tradition he began at an early age, heading up his group of school friends in their scientific and philosophical adventures. The motley crew threw ideas around between them which, while often reflecting juvenile interests, nonetheless were highly influential in focusing Hawking on those areas that truly gripped his imagination. Theirs was, in addition, a genuine meeting of minds in which strong bonds of friendship were forged. In fact, Hawking managed to keep up with most of that core group of six or seven chums throughout adulthood.

While Hawking was accused of aloofness by some of those who knew him in adulthood (an impression no doubt sometimes exaggerated because of the problems he faced in communicating), he nonetheless always proved adept at making friendships. For instance, the likes of Kip Thorne (see page 93) were undoubtedly soulmates as much as intellectual sparring partners. If the popular characterization of Hawking as 'the mind trapped in a body' conjures up images of a lonely and isolated spirit, the reality of his life serves to support John Donne's assertion that 'no man is an island'.

By the time he was embarking on his studies at Cambridge, he was under no illusion that the immense questions and enigmas he was probing could be solved by one person alone – no matter how fine a brain they might possess. As is evident from the list of collaborators below, Hawking was happy to tie his colours to the mast of whoever might help him progress his research. That, however, is not to suggest that he was interested in sycophants or only those who see the universe as he did. Take, for instance, his friendship with Martin Rees (now Baron Rees of Ludlow), a close contemporary who also studied under Dennis Sciama and who has variously served as Master of Trinity College, Cambridge, President of the Royal Society and Astronomer Royal.

The two occasionally locked horns, as in 2011 when Rees publicly took Hawking to task over comments the latter had made about the existence of God. Rees said, 'Stephen Hawking is a remarkable person whom I've known for forty years ... I know [him] well enough to know that he has read very little philosophy and even less theology, so I don't think we should attach any weight to his views on this topic.' It was quite a dressing-down for a man whom few others felt confident enough to take on – especially in public. Yet their friendship survived so that in 2015 Rees penned a celebration of Hawking's life in the *New Statesman* magazine under the title 'Stephen Hawking's life is a triumph of intellect over adversity'.

Again, the echoes of Einstein resound in Hawking's own life. While the world loves to tell stories of lone genius, the truth is usually that great intellect flourishes in company. Einstein, for instance, spent a lifetime bouncing his ideas off close friend Michele Besso and plundered the mathematical abilities of another long-term sidekick, Marcel Grossmann, when formulating his general theory. In 1902 he even established an informal salon, the *Akademie Olympia*, while he was living in Switzerland. Hawking, meanwhile, rarely shied from chewing the cosmological fat in both the private and public spheres.

There were other practical advantages for Hawking in 'staying in the swim' with his contemporaries. Not least, by keeping channels of communication open he was able to gather useful intelligence amid the highly competitive academic milieu. It was, for instance, the friendship he fostered with Jayant Narlikar that allowed him to so publicly challenge the arguments of Fred Hoyle at the Royal Academy, an act that in some senses launched Hawking into the wider public consciousness.

Hawking's office at Cambridge's Department of Applied Mathematics and Theoretical Physics (DAMTP) had a reputation for being a highly sociable place where the sharing of ideas was positively encouraged. Convention demanded that faculty members broke for morning coffee and afternoon tea, which provided convenient arenas for the dissemination of all the latest ideas and chatter. When the communication blossomed, everyone stood to benefit, from Hawking down. As he explained in a 1992 radio interview:

> I suppose I'm naturally a bit introverted ... But I was a great talker as a boy. I need discussion with other people to stimulate me. I find it a great help in my work to describe my ideas to others. Even if they don't offer any suggestions, the mere fact of having to organize my thoughts so that I can explain them to others often shows me a new way forward.

The importance of dialogue is a subject he revisited often over the years. Consider the words he spoke in another advertisement for telecommunications firm BT back in 1993 (and marvel at his ability to convey words of genuine profundity within such an incongruous context):

Mankind's greatest achievements have come about by talking, and its greatest failures by not talking. It doesn't have to be like this. Our greatest hopes could become reality in the future. With the technology at our disposal, the possibilities are unbounded. All we need to do is make sure we keep talking.

## JOINING FORCES

'He always had an amazing ability to figure things out in his head, but generally he worked with colleagues who would write a formula on a blackboard; he would stare at it, and say what should come next.'

MARTIN REES, *NEW STATESMAN*, 2015

Here, in alphabetical order, is a selection of some of the most influential collaborators from throughout Hawking's career.

**James Bardeen** (born 1939 in the USA). Having studied for his PhD under Richard Feynman at the California Institute of Technology (Caltech), in 1973 Bardeen – along with Hawking and Brandon Carter (see opposite) – set out four laws of black hole mechanics and elucidated their similarities to the long-established laws of thermodynamics. The most important of these black hole laws states that gravitational pull is equal at all points within the event horizon.

**Brandon Carter** (born 1942 in Australia). He was a fellow student of Hawking under the supervision of Dennis Sciama at Cambridge. With Hawking, Carter helped to prove the so-called 'no-hair' theory of black holes, which states that black holes are defined by the three characteristics of mass, charge and spin. Along with Hawking and James Bardeen, he also helped formulate the four laws of black hole mechanics.

**George Ellis** (born 1939 in South Africa). Like Hawking, Ellis had his PhD supervised by Dennis Sciama. Hawking and Ellis subsequently worked in partnership throughout the latter part of the 1960s and into the 1970s on singularities and fundamental questions of space-time. Their most celebrated collaboration, *The Large Scale Structure of Space-Time,* was published in 1973. It remains in print and has been described by Hawking himself as 'virtually the last word on the causal structure of space-time', although it is highly technical so he 'would caution the general reader against attempting to consult it'.

**Gary Gibbons** (born 1946 in the UK). Originally in Cambridge as another research student of Dennis

Sciama in the late 1960s, he came under Hawking's supervision in 1969. After completing his PhD on the subject of classical general relativity, he turned his attention to quantum gravity and black holes. With Hawking he developed the system of Euclidean quantum gravity that allowed for a re-imagining of time as a fourth spatial dimension. This in turn paved the way for rapid developments in the understanding of black hole thermodynamics. In more recent years he has carried out world-leading research into string theory and M-theory.

**James Hartle** (born 1939 in the USA). Long associated with the University of California at Santa Barbara, Hartle's most famous work with Hawking was conducted at the University of Chicago in 1983. It was here that the duo worked on their 'no-boundary solution', which helped redefine cosmological perceptions of the initial conditions of the Big Bang.

**Thomas Hertog** (born in 1975 in Belgium). Hertog took a physics degree at the University of Leuven in his homeland before undertaking a Masters at Cambridge. There he encountered Hawking and the pair undertook joint research into cosmic inflation. Hertog subsequently moved on to the University of California at Santa Barbara and later to CERN. It was while there in 2006 that he and Hawking published their influential model of 'top-down cosmology'. They continued to collaborate, not least on finding new practical applications for Hawking and Hartle's no-boundary model.

**Werner Israel** (born 1931 in Germany). One of his generation's leading theoreticians in the fields of gravitational physics and black holes. He co-edited two volumes with Hawking – *General Relativity: An Einstein Centenary Survey* (1979) and *300 Years of Gravitation* (1987).

**Leonard Mlodinow** (born 1954 in the USA). As well as being a respected physicist in his own right, Mlodinow is also an author and screenwriter who has collaborated with Hawking on two books: *A Briefer History of Time* (2005) and *The Grand Design* (2010).

**Don Page** (born 1948 in the USA). He was studying for his PhD at Caltech when Hawking was there in the 1970s. Hawking and Kip Thorne supervised his thesis on 'Accretion into and Emission from Black Holes'. He later moved to the University of Cambridge, where he worked as a research assistant to Hawking. Over the following decades they worked together on several academic papers on subjects including entropy, wormholes, inflation, black hole thermodynamics and quantum gravity.

**Malcolm Perry** (born 1951 in the UK). Like Hawking, Perry took his undergraduate degree at Oxford before moving to Cambridge. He was then under Hawking's wing from the mid- to late 1970s, receiving a PhD for his work on black holes and quantum theory. He subsequently collaborated with Hawking and Gary Gibbons, contributing to the development of Euclidean quantum gravity and carrying out research into black hole radiation. He then spent several years at Princeton University in the USA but returned to Cambridge in 1986 and has remained there ever since, becoming a world authority in string theory and M-theory. He and Hawking continued to work together academically and Perry also serves as a trustee of the Stephen Hawking Foundation. (The Foundation was launched in 2015 with the triple aims of furthering research into cosmology, astrophysics and fundamental particle physics at school and university levels; facilitating those working with ALS; and researching and contributing to the better care of those with ALS.)

**Roger Penrose** (born 1931 in the UK). Having received his doctorate from the University of Cambridge in 1957, Penrose gave the lecture seven years later that in certain respects turned around the academic fortunes of the young Hawking. The two men subsequently worked together on black hole singularities and in the 1970s reported a series of landmark theories seeking to explain the relationship between singularities and gravitational collapse in Einstein's general theory. The two shared numerous international awards for their work, as well as co-authoring *The Nature of Space and Time* (1996) and *The Large, the Small and the Human Mind* (with Abner Shimony and Nancy Cartwright, 1997). However, Penrose also critiqued some of Hawking's work, not least his championing of M-theory, which he has described as 'hardly science' but rather a 'collection of hopes, ideas and aspirations'.

**John Stewart** born in the UK (1943–2016). Stewart, a doctoral student of George Ellis (see page 89), was a colleague of Hawking at DAMTP. Together they produced a celebrated paper in 1993 that introduced the idea of 'thunderbolt singularities'. The pair set out to investigate the nature of so-called naked singularities – in other words, singularities that exist without an event horizon. The question of whether naked singularities can exist in reality remains a controversial one for many scientists. Nonetheless, Stewart and Hawking came to the startling conclusion that in theory certain naked singularities (nicknamed thunderbolt singularities) might start rapidly expanding at the speed of light from the instant of their birth, destroying everything in their path.

**Kip Thorne** (born 1940 in the USA). One of the world's leading astrophysicists, Thorne first worked alongside Hawking at Caltech in the mid-1970s. Indeed, the opportunity to do so was one of the leading attractions for Hawking to move to Caltech. In 1975, the pair undertook one of the most famous wagers in modern science. A decade or so earlier, an astronomical phenomenon known as Cygnus X-1 was discovered in the Cygnus galaxy, emitting X-rays that suggested it might be a black hole. Thorne held that it was, while Hawking bet it wasn't. It was a hedge on Hawking's part, since he sincerely hoped it was but figured that if proved not to be, he could at least enjoy getting one over on his pal. Hawking conceded the bet in 1990 (to both men's satisfaction), presenting Thorne with a subscription to *Penthouse* magazine as his prize (to Mrs Thorne's reported dissatisfaction). Their friendship continued to thrive, with Thorne a trustee of the Stephen Hawking Foundation.

**Neil Turok** (born 1958 in South Africa). Director of the Perimeter Institute for Theoretical Physics, based in Ontario, Canada from 2008 to 2019, his work with Hawking in the 1990s resulted in the Hawking–Turok instanton theory and a model that allows for inflation within an open universe structure. Like Thorne, Turok also entered into a famous wager with Hawking. In 2002, Hawking bet him that cosmologists would discover gravitation waves proving the theory of inflation as correct. The bet seemed to have been settled in Hawking's favour when in September 2015 the Laser Inferometer Gravitational-Wave Observatory (LIGO) recorded such waves for the first time.

# Even a Brilliant Intellect Needs Emotional Succour

'I couldn't carry on with my life if I only had physics. Like everyone else, I need warmth, love and affection.'

STEPHEN HAWKING, *BLACK HOLES AND BABY UNIVERSES AND OTHER ESSAYS*, 1993

It was easy to imagine Hawking as an almost 'other-worldly' figure, sitting in his chair day and night, processing enormously long equations and mapping corners of the universe that most of us couldn't begin to comprehend. But the image of Hawking the scientist should not obscure that of Hawking the human being. Much more than merely the *boffin extraordinaire* of his age, he was a remarkably rounded figure, entertaining a wide range of passions and interests, as well as struggling with the full gamut of emotional ups and downs to which all of us can relate. Hawking was a major public figure because of his contributions to physics, but his achievements in the discipline represent, let it not be forgotten, but a part of the man.

We have already looked at the important place music had in his heart and, as we shall go on to see, he also took a keen interest in everything from literature and politics to movies and television. He enjoyed some sport, too, being a keen cox and cyclist before disability robbed him of those pleasures, and he also spoke of his love of food and drink (sherry was a favourite tipple as a young man, while curry, chocolate truffles and, above all, crème brûlée were guaranteed to stimulate his taste buds). His wide-ranging interests suggested a character who was emotionally responsive to the world around him and who craved pleasure beyond that which he could obtain from his work.

While a beautiful piece of music and a box of chocolates can sometimes be just what a weary soul needs, there is nothing that quite matches the emotional succour one can gain from interpersonal relationships. For all that Hawking was married to his work, he also entertained a rich and sometimes tempestuous personal life. Among his many cosmological insights, we might assume that he concluded long ago that love makes the world go round.

Indeed, of much more lasting importance than any temporary sensual pleasure were Hawking's two marriages. The transition from the first (to Jane Wilde) to the second (to Elaine Mason), and then the collapse of that second marriage too, brought a level of scrutiny to his private life that he was neither prepared for nor happy about. These relationships are looked at in more detail on pages 100 to 105; suffice it to say that he came to cite them both (and the family that resulted from the first marriage) as achievements in his life that ranked alongside his work in cosmology.

## A SUCKER FOR A PRETTY FACE

In countless interviews, Hawking was happy to acknowledge his love for women, while recognizing that they remained an utter enigma to him. In crude terms, he was always a sucker for a pretty face, ever since Marilyn Monroe sauntered into his consciousness as an adolescent (a Marilyn lookalike was even employed to cut the cake at his sixtieth birthday celebrations). In more recent years, it was reported that he enjoyed the attentions of comely hostesses at certain London nightspots, with infamous nightclub owner Peter Stringfellow commenting in 2011: 'He's a man who lives within his brain and still manages to feel the over-whelming power of sex.'

In later years Hawking lived alone and employed a housekeeper to look after his domestic affairs. But he built bridges with the family from his first marriage and had a much calmer and more settled private life. This part of his life was not always a happy one and his conduct sometimes hurt those closest to him. While he achieved so much, his loved ones were sometimes required to make extraordinary sacrifices for him. Nonetheless, there is something admirable about the way in which Hawking

– whose physical impediments were so oppressive and whose name is so widely associated with reason and rationality – allowed passion (with all its positive and negative connotations) to weave itself through the fabric of his life.

When Hawking was asked (as he regularly was by interviewers) if there was any mystery that continued to baffle him, his go-to answer was 'women'. They were, he confided to *New Scientist* in 2012, the subject he spends most time thinking about. While Hawking will live on as a giant of science, it is comforting to know that an intellect as sophisticated as his required the same sort of emotional nourishment as the rest of us. That he was as prone to making missteps in his relationships as anyone else suggests a human frailty that too often goes unacknowledged.

## A BRIEF HISTORY OF MARRIAGE AND DIVORCE

> 'Without the help that Jane has given I would not have been able to carry on, nor have the will to do so.'
>
> STEPHEN HAWKING, QUOTED IN *STEPHEN HAWKING: HIS LIFE AND WORK*, KITTY FERGUSON, 2011

Hawking was not afraid of making bold claims, so perhaps we should not fear doing the same. One with which he would doubtless have concurred is that without the efforts of his first wife, Jane, Hawking could not have built the career that he enjoyed.

The couple met for the first time at a New Year's Eve party to celebrate the arrival of 1963. This was before Hawking's diagnosis of ALS but after the first symptoms had started to show themselves. She was younger than him, still doing her A levels and planning to study modern languages at Westfield College, University of London. Smart and self-assured, she was attracted not only by Hawking's obvious intelligence but also by the sense of there being 'something lost' about him.

Their relationship quickly blossomed and Hawking found that being part of a team and having someone else who relied upon him spurred him on in his work. However, his new-found career focus was not without its price. In order to make significant strides in a discipline – black hole cosmology – that was then in its infancy required unstinting dedication. If his motives to make the best of himself were in part selfless, the realities of his day-to-day labours required a degree of self-absorption. Furthermore, he was finding that he actually *enjoyed* throwing himself into his work.

In those early days, Jane was happy to go along with the situation. Having completed her schooling, she began her degree in London and treasured the time she got to spend with Hawking at weekends and in the holidays. Furthermore, Hawking's diagnosis had left them both under the impression that their life together would likely be short. Every moment was of immense value and she was intent to cherish them all.

There were early warning signs, however, of problems ahead. For instance, in 1965 Hawking decided to apply for a fellowship at Cambridge's Gonville and Caius College. He and Jane had become engaged and the position would give them a semblance of security. However, he was already struggling to write by then, so lined up Jane to complete the application. The week before she was due to visit and carry out the task, she broke her arm. Hawking was, by his own admission, distinctly lacking in sympathy, instead dwelling on the repercussions for himself. In fact, Jane managed to handwrite the application before another friend typed it up and Hawking, needless to say, was awarded the fellowship.

The pair married that year and their first child, Robert, was born in 1967. A girl, Lucy, followed three years later. Soon the strain began to tell on Jane. Not only did she have two young children in her charge, but her husband's own care requirements were rapidly increasing, too. They did get extra help in the shape of visiting nurses and then live-in academic assistants who also undertook some domestic tasks for Hawking. However, she missed the privacy they had once enjoyed and began to feel like a spare wheel as Stephen's career went into overdrive. The arrival of a third child, Timothy, in 1979 only added to her workload even as he was a much-loved addition to the family.

As she would tell the *Telegraph* in 2015: 'Sometimes life was just so dreadful, so physically and mentally exhausting, that I wanted to throw myself in the river – although of course I stopped myself because of the children.' In an interview in the same year with the *Guardian*, she expanded on the difficulties she faced: 'The truth was, there were four partners in our marriage. Stephen and me, motor neurone disease and physics. If you took out motor neurone disease, you are still left with physics.' She also pointed to another echo of Einstein in her husband's life: 'Mrs Einstein, you know, cited physics as a [reason] for her divorce ...'

While his star rose throughout the 1980s, Jane felt increasingly trapped in her existence and distant from her celebrated husband. 'Fame and fortune muddied the waters and really took him way out of the orbit of our family,' she told the *Observer* in 2004. It was a sentiment her daughter, Lucy, put rather more bluntly, as quoted in a *Vanity Fair* article in the same year: 'The whole of my early life I looked after him, when he wasn't rich and he wasn't famous, and we all did – because we loved him. And the minute he got fame and money he was gone.' In Jane's words, her husband had become an 'all-powerful emperor' and a 'masterly puppeteer'.

By the late 1980s the marriage had effectively broken down. Following Tim's birth, Jane had invited an organist from the local church, Jonathan Hellyer Jones, to move into the Hawking household. For

Jane, it was something akin to an insurance policy should anything happen to Stephen. She holds that the arrangement was initially platonic and that her husband was tacitly accepting of it. He, however, suggested he felt unable to object but resented their growing closeness. Meanwhile, he found himself falling for one of his nurses, Elaine Mason, who was married to the man who had helped Hawking make various adaptations to his computers so that he could operate them from his chair.

In 1990 Hawking called time on the marriage and moved in with Elaine. Stephen and Jane divorced in 1991 and he married Elaine four years later. Jane likewise wed Jonathan in 1995. The former Mr and Mrs Hawking were barely on speaking terms and his relationship with the children was hardly better. Meanwhile, the changing sands of his domestic set-up captured the imagination of the popular press. Suddenly, the revered genius found his personal life was tabloid fodder.

Inherently protective of his privacy, Hawking found the intrusion hard to bear, but worse was to come. In 2000 he and Elaine were at the centre of a police investigation into allegations – backed in the press by statements from unnamed nursing staff – that his wife was abusing him. His daughter Lucy was one of those who contacted police demanding they look into an apparent series of unexplained physical injuries, not to mention reports of emotional abuse.

Hawking was incandescent that his personal life was being played out under the glare of publicity. 'I firmly and wholeheartedly reject the allegations that I have been assaulted,' he said in a 2000 statement. Nor could the police find evidence to substantiate the accusations. In the end, he and Elaine divorced in 2006 but he never suggested she was guilty of any abuse. Their relationship, he said, was 'passionate and tempestuous' and had its 'ups and downs', yet he also thanked her for saving his life on several occasions.

Following the divorce from Elaine, relations between Stephen and Jane and his children thawed considerably (so much so that he and Lucy co-wrote a series of books; see 'Write Like Hawking', page 141). This is reflected in Jane's 2007 book about their life together, *Travelling to Infinity: My Life With Stephen*, upon which the movie *The Theory of Everything* was based. It was a revision of an earlier book, 1999's *Music to Move the Stars: A Life With Stephen*, and much of the rawness evident in that work had clearly subsided. As the *Independent on Sunday* described it: 'This is not a vindictive book, although the agony she went through is palpable; if Stephen's struggle to keep his mind clear is heroic, so is her determination to balance his escalating needs and those of their three children.'

It may not have always been pretty, but Hawking's marital history reflects a complex and rich life that further belies the popular notion of the calm genius within a crippled body.

# Careers Are Made, Not Born

'Science is a scrum. Most of us are working in the scrum. Stephen, by the very nature of his disability, can't work in the scrum.'

BERNARD CARR, PROFESSOR OF MATHEMATICS AND ASTRONOMY AT QUEEN MARP, THE UNIVERSITY OF LONDON, AND A FORMER PHD STUDENT UNDER HAWKING, INTERVIEWED IN THE *GUARDIAN*, 2008

While the debate rages as to exactly how the universe came into being, we can be more certain of one thing: careers *do not* spontaneously occur. And they certainly don't undergo phases of impromptu rapid expansion. A successful career requires several ingredients – a smattering of natural talent, a good work ethic and a fortuitous alignment of the stars, for example. But equally important is a bit of strategy.

The young Hawking harboured ambitions of doing extraordinary things, but it was only with the approaching end of his time at Oxford (and, of course, the arrival on the scene of Jane Wilde) that he seems to have got really serious. When it came to choosing his PhD, he was drawn to cosmology not only because he was instinctively interested in it, but also because he saw that it was a hitherto neglected area of research on the cusp of remarkable breakthroughs. He sensed there was about to be a cosmological party and he wanted to be on the guest list. In choosing where to study, he was most certainly influenced by having the opportunity to work under the (at the time) biggest name in the field, Fred Hoyle. That fate would step in to ensure he was placed, instead, with Dennis Sciama (someone, it would turn out, with whom he far more closely related) is perhaps an example of how success also requires a dose of good luck.

An instinct for working in the right places with the right people was a hallmark of Hawking's career. As is evident from the earlier survey of his many collaborations, he worked with a great many of the biggest names in physics of the last sixty years, starting with Sciama and Roger Penrose. That was not something that happened only by chance, but was the result of considered strategy. He kept himself surrounded by other bright intellects who he knew would bring out the best in him, just as he pushed them on, too.

Similarly, while DAMTP in Cambridge provided him with his spiritual home for the large part of his working life, he had stints at a number of other major institutions when it suited his overarching career aims. In 1974, for instance, he relocated his young family to California in order to take up the chance of studying at Caltech, funded by a Sherman Fairchild Distinguished Scholarship. There he would mix with such luminaries as Richard Feynman and Kip Thorne. It was a bold move but one he sensed he had to take. He was proved spectacularly correct.

Caltech provided him with the care he then needed

in light of his failing health (arguably the institution was in a much better state to do so at the time than the authorities in Cambridge) and gave his family the chance to taste the American Dream for a while. More importantly, his stay helped him consolidate his growing reputation on the international academic stage. Caltech, after all, has produced over thirty Nobel laureates despite its relatively small faculty size. On his arrival back in Cambridge, he was received as a returning hero. The practical impacts of this included improved domestic arrangements, with the university assisting in finding his family grander and more suitable accommodation.

Hawking was always sanguine when it came to addressing his personal needs so that he might work to the greatest effect. He accepted help as he required it, speaking out when it was unforthcoming and refusing to martyr himself by suffering in silence. His achievements, extraordinary as they are, would have been utterly impossible if he had not sought out support wherever it was available. This was most clearly seen in his willingness to utilize the latest technology in order to make daily living easier, from the computer adapted to his wheelchair that allowed him to work electronic devices to the voice synthesizer that helped define his public persona. A greater personal cost came with his acceptance of live-in help from the mid-1970s onwards, as his family became unable to care for him on their own.

Hawking characteristically 'got on with it', employing a succession of research students to assist him, beginning with Bernard Carr. They helped him with the nitty-gritty tasks of day-to-day life in return, as Carr put it, for 'participating in history'. So Hawking transformed an incredibly delicate situation into a win–win scenario.

Although he stayed only a year in California, Hawking maintained close ties with the institute and its world-leading researchers. He also kept a hand in at other cutting-edge institutions, including the European Organization for Nuclear Research (CERN) in Switzerland. Home to the Large Hadron Collider, it is arguably currently the world's most important centre for particle physics research and, although Hawking was never on its staff, he paid frequent visits to keep abreast of the latest ideas emanating from there. Meanwhile, in 2008 he accepted the position of Distinguished Visiting Research Chair at the Perimeter Institute for Theoretical Physics in Ontario, Canada. An independent research facility, it is now the largest institute in the world focused exclusively on theoretical physics. In 2011 it opened a major new complex, which it named in honour of Hawking, who had several stays on site.

As for Cambridge, it did its bit to ensure that Hawking felt valued by his alma mater. Most significantly, he joined the prestigious list of Cambridge's Lucasian Professors of Mathematics, a position inaugurated in 1663. Hawking was just thirty-seven when he acceded to the famous chair and he held the post for thirty years, retiring in 2009 aged sixty-seven as tradition demands. He was Director of Research at DAMTP until 2014, when he became its inaugural Chair of Cosmology, and also headed Cambridge's Centre for Theoretical Cosmology, a department he founded in 2007. Cambridge bestowed another major honour upon him in 1989 when it made him an honorary Doctor of Science.

Hawking was undoubtedly a smooth operator, effectively climbing the greasy pole of academia and clinging on at its apex for decades. Although he spent many years living on a modest academic salary, he was also known as a wily negotiator who rarely undersold himself. He expected, justifiably, fair remuneration for his unique expertise and ever since the 1980s, when his fame began to escalate, he commanded top dollar for writing and speaking engagements. Furthermore, he became a master of publicity. He knew that his public utterances would garner plentiful attention and kept his profile high by happily courting a little controversy now and again.

For instance, he contributed to debates on everything from the likelihood of time travel to the dangers of artificial intelligence and the existence of alien life. On occasion, his outspokenness drew criticism from fellow scientists, who variously questioned the extent of his expertise in certain areas and the undue influence his words sometimes seemed to carry. But Hawking continued to work the media unabashed. There was a reason he became the world's most famous living scientist, after all. While his exact wealth is unknown, it is believed to run into several million pounds. There are not many theoretical physicists who could say the same.

## THE NOBEL SNUB?

'I think most theoretical physicists would agree that my prediction of quantum emission from black holes is correct, though it has not so far earned me a Nobel Prize because it is very difficult to verify experimentally.'

**STEPHEN HAWKING,** *MY BRIEF HISTORY*, **2013**

Hawking was not short of accolades and awards throughout his career. They include, for instance, such prestigious gongs as the Albert Einstein Award, the Dirac Prize and the Fundamental Physics Prize. The latter, which was given to Hawking in 2013, was founded by Russian scientist and businessman, Yuri Milner, for contributions to fundamental research. Worth a cool US$3 million, it is financially the most valuable academic prize in the world. Other honours bestowed upon Hawking include election to the fellowship of the Royal Society when still in his thirties, and his appointment as a Commander of the Order of the British Empire in 1982. He was also the recipient of the Presidential Medal of Freedom, the highest civilian award in the United States.

Yet one prize is conspicuous by its absence from his CV – that of the Nobel Prize for Physics. At first glance, it feels odd that one of science's most celebrated figures should be deprived of its greatest tribute. Was he the victim of some conspiracy of jealousy or prejudice? Or was it the case, dare we even ask, that he wasn't *quite* brilliant enough to warrant it? The answer on both counts is 'No'. But for Hawking to get a Nobel Prize, certain unlikely criteria would have needed to be met.

As will be seen in the section on 'Hawking's Legacy' at the end of this book, there are those who wonder whether his abilities as a theoretician were overestimated. It is true that there are a number of extravagantly gifted scientists working in his field who have made astonishing breakthroughs of their own, but who nonetheless lack Hawking's profile and so – some suggest – don't get the credit they deserve by comparison. However, few disagree that Hawking held his own among the planet's scientific elite and all but the most grudging concede that he should have been considered a contender for the Nobel Prize. As Brian Cox, the celebrated broadcaster and professor of particle physics at Manchester University, has put it, he was 'a genius of Nobel-Prize calibre'.

The Nobel Prize for Physics has been awarded since 1901 in accordance with the rules stipulated in the will of Alfred Nobel, the Swede who invented dynamite.

Nobel feared being remembered primarily as the creator of an instrument of destruction, so intended the awards to celebrate those who most positively contribute to the world. The physics prize was, therefore, to be given to 'the person who shall have made the most important discovery or invention within the field of physics …' At the time of writing, over two hundred Physics laureates have been created over the prize's lifespan, but not Hawking, despite the existence of a pro-Hawking lobby within the Nobel Academy for well over thirty years.

It is generally acknowledged that Hawking was denied a prize because his most celebrated work – concerning the existence of Hawking radiation – cannot be proved by experiment. While his thesis was widely accepted and incorporated into theoretical models, no one has been able to actually witness the phenomenon. Whether we will ever manage to study a black hole in the process of evaporating remains moot, but without such an evidential basis the Nobel Prize continued to elude him.

Hawking acknowledged as much, although he still held out hope for a scientific miracle in his favour. As he told journalists in 2008 in reference to the newly operational Large Hadron Collider, 'Most exciting for me would be if it found little black holes, for then I would get the Nobel Prize. However, I don't think it is very likely, and I'm not holding my breath.'

Hawking was by no means the first theoretician to

miss out on the grandest prize of them all. No lesser light than Einstein almost fell victim to the same plight, before finally winning one in 1922. While most people accepted that the general theory of relativity was his outstanding contribution to mankind, the Nobel committee faced the same problem as they did with Hawking: no one could prove the theory by observable experiment. Therefore it remained a 'theory', and not a 'law' that could meet Nobel's demand for a 'discovery or invention'. The committee's answer was, ultimately, to fudge around the issue. Einstein actually received the Prize – twelve years after being first nominated – for 'his services to theoretical physics, and especially for his discovery of the law of the photoelectric effect'. In other words, he got the prize for a much less celebrated, brilliant and influential work than the general theory.

Just as Marcel Brillouin, who proposed Einstein for the award in 1922, pointed out how it would seem fifty years down the line if Einstein continued to be overlooked, we may wonder the same about Hawking. The man himself continued to play it cool, though. As he commented when receiving the Fundamental Physics Prize (worth more than twice as much as the Nobel Prize): 'No one undertakes research in physics with the intention of winning a prize. It is the joy of discovering something no one knew before …'

# Getting It Wrong
# is Not a Sin

'Some people will be very disappointed if there is not an ultimate theory that can be formulated as a finite number of principles. I used to belong to that camp, but I have changed my mind.'

STEPHEN HAWKING, 'GÖDEL AND THE END OF PHYSICS' LECTURE, 2002

As explored in the earlier chapter 'Have No Limit to Your Ambition', Hawking's position on a theory of everything did fundamentally change. From believing that the human race was in touching distance of discovering it, latterly he suggested that such an ambition was likely unattainable. Such sea changes have a habit of disconcerting observers, in the same way that voters are often turned off by politicians who flip-flop on policy.

There is a breed of politician who adapt their politics in order to bend to whichever way the wind is blowing and they are rightly deserving of distrust. But there is another type who changes their mind in the light of new evidence or after serious reconsideration of all the information to hand. Hawking did a similar thing in scientific terms, so that the act of changing his mind may be seen as admirable, and even brave. Where he could hide behind his reputation as a 'genius', he instead exposed himself to scrutiny (not to mention potential ridicule) by admitting that he took an intellectual misstep. He did so in the knowledge that only by conceding and learning from our mistakes can we make true progress.

By Hawking's own estimation, the biggest about-face of his career concerned what is known as the black hole information paradox. In this context, 'information' is the data that defines the properties of a given object – for instance, its size, shape, colour or charge. According to the no-hair theory to which Hawking adhered, black holes have very low information content (that is to say, they are described only by their mass, spin and charge). Yet we know that anything that falls into a black hole is unable to escape back out, so what happens to all the unique information attached to those objects? Hawking's long-held belief was that it is destroyed within the black hole but this seemed to contravene other fundamental scientific laws that state such information must be conserved – hence the paradox.

The question of the fate of information in a black hole became the subject of a famous bet in 1997, between Hawking and Kip Thorne on one side (holding that it is destroyed) and their fellow Caltech alumnus, John Preskill, on the other (who argued that Hawking radiation permits black holes to 'leak' information). In 2004, Hawking admitted defeat on the basis of the most of-the-moment quantum thinking, presenting Preskill with an encyclopaedia of baseball, a symbolic source from which 'information may be freely accessed'. (For the record, Thorne has yet to give up on the bet.)

In 2012, Hawking told the *New Statesman*:

> I used to think that information was destroyed in black holes. But the AdS/CFT correspondence [anti-de Sitter/conformal field theory correspondence; an implied relationship between separate theories encompassing quantum field theories and string theory] led me to change my mind. This was my biggest blunder, or at least my biggest blunder in science.

Seeking to make good on his perceived mistake, in 2015 he announced a forthcoming paper he had written that was widely trailed as providing 'the answer to the black hole information paradox'. In the article he posits that particles entering a black hole may create disruption that results in information being imprinted on outgoing Hawking radiation. While the media went into an inevitable frenzy, most experts in the field argued that, while an interesting hypothesis, it was some way off providing a definitive answer to the paradox. It was nonetheless another staging post in the complex recalibration of his thoughts on the subject.

Nor is it the only major U-turn he made. In terms of high-profile climbdowns, his change of heart over the much fabled Higgs boson ranks highly. Named after the British physicist Peter Higgs, the Higgs boson is a sub-atomic particle, the existence of which was first proposed in 1964. It is, in short, a mechanism that can be used to explain how certain particles gain mass, a question hitherto unanswered in the widely accepted standard model of elementary particles. In the early 2000s, Higgs and Hawking embarked on a heated – and not always edifying – public debate as to its existence, with Hawking insistent it would not be found.

However, the Higgs boson was discovered at CERN's Large Hadron Collider in 2012. Hawking was wise enough to see that the game was up, even pushing Higgs forward for a Nobel Prize. Higgs was duly honoured by the Nobel Academy in 2013, sharing the award with François Englert. Hawking was gracious in his praise even as he was amusingly sardonic:

A few weeks ago, Peter Higgs and François Englert shared the Nobel Prize for their work on the boson and they richly deserved it. Congratulations to them both. But the discovery of the new particle came at a personal cost. I had a bet with Gordon Kane of Michigan University that the Higgs particle wouldn't be found. The Nobel Prize cost me $100.

Nor could he resist a final wry dig at his rival, reflecting on the Higgs boson: 'Physics would be far more interesting if it had not been found.'

# Getting It Wrong is Not a Sin

Ultimately, Hawking was consistently open to the chance that he might be wrong on any given subject. For all the bluster that occasionally accompanied his public pronouncements, he never lost sight of the precariousness of the theoretical physicist's trade. As he put it in *My Brief History*:

Any physical theory is always provisional, in the sense that it is only a hypothesis: you can never prove it. No matter how many times the results of experiments agree with some theory, you can never be sure that the next time the result will not contradict the theory. On the other hand, you can disprove a theory by finding even a single observation that disagrees with the predictions of the theory.

# Read Like Hawking

'Science fiction … is not only good fun but it also serves a serious purpose, that of expanding the human imagination.'

STEPHEN HAWKING IN THE PREFACE TO LAWRENCE M. KRAUSS'S *THE PHYSICS OF STAR TREK*, 1995

It is safe to say that Hawking was remarkably well read in the literature of science as it related to his particular areas of interest. From Copernicus, Galileo, Kepler and Newton via the likes of Einstein, Dirac and Planck, on to the works of his many notable contemporaries, Hawking immersed himself in his subject matter. This in itself was an extraordinary achievement, given that even the simple act of accessing literature drained his physical resources in a way that few of us, thank-fully, can relate to. For instance, Martin Rees has recalled how challenging it was for Hawking to read a book as far back as the 1970s. He tells of how he would sometimes push Hawking's wheelchair back to his office, where Hawking would ask him to leave a particular tome open at a specific point. 'He would sit hunched motionless for hours,' Rees said. 'He couldn't even turn the pages without help.'

Two of Hawking's own books provide a good starting place from which to investigate some of the scientific works that most influenced him – 2002's

*On the Shoulders of Giants* (see page 34) and 2005's *God Created the Integers*, outlining his choice of the greatest mathematical works in history from Euclid to Alan Turing.

But Hawking did not restrict himself only to works of non-fiction. He was, for instance, a great champion of science fiction – a genre in which he read extensively as a teen (having been something of a latecomer to reading itself). The attraction was obvious, as the genre gave him the chance to combine his love for science and technology with his fascination for what lies beyond our own planet. In other words, a good slice of science fiction allowed for the imaginative contemplation of the types of existential questions that were meat and drink to him.

He was lucky enough to live through a golden age of science fiction. Such legendary names as Isaac Asimov, Philip K. Dick, Arthur C. Clarke and Kurt Vonnegut were all in their prime as Hawking let loose in the genre. This was in addition, of course, to all those works that had already achieved classic status by then – novels like H. G. Wells's epic of time travel, *The Time Machine*. One can imagine the young Hawking fixated as he read that book, which some have argued presaged the general theory of relativity by a good twenty years. As Lawrence Krauss put it in his critical work, *Hiding in the Mirror*, 'the very first page … included an explanation from the unnamed time traveller about how objects require existence in time as well as space. To modern ears, his description sounds a lot like Einstein's vision of space and time.'

Hawking also eagerly devoured the works of Aldous Huxley (1894–1963), an English writer whose novels transcended traditional genre boundaries. His most celebrated work is *Brave New World*, published in 1932 and presenting a dystopian vision of London set several centuries into the future. Taking in themes as disparate as advances in reproductive technology and psychological conditioning, he depicted a frightening world that posed serious philosophical questions for its readers. This was truly science fiction to make you think. However, Hawking was not a fan of the author's entire output. For instance, he read one of Huxley's earlier novels, 1928's *Point Counter Point*, while he was a student at Oxford and found the long (it is the longest of all Huxley's works) and complex multi-stranded narrative rather a bore. It is a 'very bad novel', he argued, a judgment which reveals that Hawking was happy to wield the critical scalpel and was no respecter of reputation for its own sake.

Always a discerning consumer of the genre, in 1992 Hawking said, 'I read a lot of science fiction when I was a teenager. But now that I work in the field myself, I find most science fiction a bit facile … Real science is much more exciting because it is actually happening out there. Science-fiction writers never suggested black

**SCI-FI AND THE SCIENTIST**

Hawking's fascination with sci-fi did not wither with advancing years, as he sated his appetite not just with books but with films and television shows too. He was, for instance, a professed fan of *Star Trek*. In his preface to Krauss's celebrated investigation into the science of that show, Hawking expounded on why he continued to be drawn to high-quality science fiction. 'We can explore how the human spirit might respond to future developments in science,' he wrote, 'and we can speculate on what those developments might be.'

Meanwhile, in 2007 he introduced a series of one-hour sci-fi dramas for the ABC channel called *Masters of Science Fiction*. Each episode adapted a story by such respected practitioners as Harlan Ellison, Howard Fast, Robert A. Heinlein, John Kessel, Walter Mosley and Robert Sheckley.

holes before physicists thought of them.' And in an interview with satirist and broadcaster John Oliver in 2014, he even sent out a challenge to the current generation of sci-fi authors – to incorporate imaginary time into their plots: 'It's the one bit of my work that science fiction writers haven't used because they don't understand it.'

While Hawking wore his sci-fi geek credentials with pride, his literary tastes were pretty catholic. Among the authors he tried in his younger days, for instance, was Kingsley Amis (1922–95). While Amis eventually came to garner a reputation as something of a curmudgeon with distinctly questionable personal politics, when he burst on to the literary scene in 1954 with *Lucky Jim*, he was regarded as a pivotal member of the so-called Angry Young Men movement. Amis, like most of the writers associated with the group (John Osborne, John Braine, John Wain and Alan Sillitoe among them), tried to distance himself from the tag. Nonetheless, this loose affiliation of working-class and middle-class writers – railing against the inequities of post-war British life under the rule of the Establishment – no doubt appealed to Hawking's own feelings of disillusionment and ennui at this time. Furthermore, *Lucky Jim* rates as among the best comic novels in the language; its tale of the misadventures of Jim Dixon (a young, disgruntled lecturer in medieval history at a provincial university) would have appealed to Hawking's impish sense of fun. It might also be suggested that Hawking and Amis harboured lifelong bewilderment as to the nature of women.

We know, too, that he was a fan of William Golding, (1911–93), who was awarded the Nobel Prize for Literature in 1983 and who is best known for the novel *Lord of the Flies* (published in the same year as *Lucky Jim*).

The story of a group of boys stranded on an uninhabited island who attempt to live under their own system of government, it was one of the most celebrated books of the century for its study of human nature and the battle between personal interest and common good. Dystopian in tone, it is also highly political and speaks of Hawking's own investment in issues beyond the lecture theatre. So too his interest in Bertrand Russell (1872–1970), the British philosopher, mathematician and all-round polymath. A logician of the first order – who, in his own words, 'imagined myself in turn a Liberal, a Socialist, or a Pacifist' – Russell rarely if ever shied from an intellectual challenge. Awarded the 1950 Nobel Prize for Literature 'in recognition of his varied and significant writings in which he champions humanitarian ideals and freedom of thought', he was just the sort of 'big ideas guy' with whom Hawking could get on board.

However, in his 1992 appearance on *Desert Island Discs*, Hawking nominated one of the great Victorian novels as the book that he would wish to be shipwrecked with (along with the *Complete Works of Shakespeare* and the Bible, the default titles all castaways are given by the show's producers). His choice was *Middlemarch* by George Eliot (pseudonym of Mary Ann Evans, 1819–80), which he described as 'a book for adults'. Originally published in instalments during 1871 and 1872, it is a masterpiece of characterization and plotting. With a large cast weaving their way through several interconnected plots, it ruminates on profound questions of politics and philosophy that remain as pertinent today as when it was first written. In a sense, it captures a whole world within its pages and it is perhaps the opportunity it offers to cast one's mind over an entire captive literary cosmos that appealed to Hawking.

# Stand Up for What You Believe in

'Nuclear war remains the greatest danger
to the survival of the human race.'

STEPHEN HAWKING, SPEAKING AGAINST THE
BRITISH GOVERNMENT'S RENEWAL OF THE
TRIDENT NUCLEAR DEFENCE SYSTEM, 2011

Given his achievements in the face of great adversity, Hawking became imbued with an air of moral authority that few public figures could rival. Allied to the fact that there were but a small number of people bold enough to consider themselves his intellectual equal, he thus found himself in a virtually unique position from where he could influence public opinion and government policies. It was a task he embraced with vigour.

He was brought up amid an atmosphere of political activism, especially on his mother's side. She was, for instance, once a member of the British Communist Party as well as an early and committed member of the CND peace movement. Hawking no doubt also became aware at an early age that one of his heroes, Albert Einstein, had used his fame as a scientist to draw attention to wider political and social issues with enormous effect.

Just as Einstein campaigned to promote nuclear

disarmament – a technology his scientific breakthroughs had made possible, to his horror – so Hawking took on the mantle. In 1946, Einstein wrote in *The New York Times*: 'The unleashed power of the atom has changed everything save our modes of thinking and we thus drift toward unparalleled catastrophe.' What he would have made of Hawking – his successor in the public consciousness – needing to press home the same message more than sixty years later we can only guess. Hawking also came out as a vocal opponent of the 2003 invasion of Iraq, which he described as a war crime on the grounds that the decision to take military action was 'based on lies'.

However, it is perhaps as a campaigner for disability rights that Hawking made the greatest impact. As quite probably the most famous disabled person on the planet, when he spoke on the subject, others were compelled to listen. As long ago as 1979 he was named Man of the Year by the Royal Association for Disability and Rehabilitation. His quest to improve conditions for disabled people began in earnest during that decade, when he waged his own personal battle with the authorities at Cambridge. In that different age, he found the University rules inflexible when it came to, for instance, providing him with housing that catered to his particular needs, or adapting the DAMTP building to accommodate his wheelchair.

In a speech he gave at Gonville and Caius College

in 2015, he painted a sympathetic picture of the help he has received over the years: 'A flat was adapted for me and my family,' he said. 'The college installed a lift in its beautiful medieval buildings, and they were adapted for the twenty-first-century technology I need to get around and work.' However, this was a rather rose-tinted recollection, leaving out much of the gory detail of the pressure needed to push the college into action. It was, we may assume, a tactic designed to hammer home the point he really wanted to make that evening, as he highlighted deficiencies in higher-education budgets:

> I wonder whether a young ambitious academic, with my kind of severe condition now, would find the same generosity and support in much of higher education. Even with the best goodwill, would the money still be there. I fear not.

In 2006 his quest for better disabled access went international when he fronted an advertising campaign promoting disability rights in Israel. 'In twenty years, we may live on the moon,' his famous voice intoned, continuing, 'during the next two hundred years we may leave the solar system and head for the stars, but meanwhile we [the wheelchair-bound] would like to go to the supermarket, cinema and restaurants.' Six years earlier he had teamed up with eleven other prominent global figures, including Archbishop Desmond Tutu, in a call to arms to work towards disability prevention and protect disabled rights. The movement's focal point was the Charter for the Third Millennium, which demanded that governments demonstrate the political will to prevent easily avoidable conditions and illnesses which cause disability. Hawking was also a staunch defender of the right to use stem cells in medical research (although he cautioned against expecting a slew of 'miracle cures') – a controversial area of science, which he claimed was being stifled by 'reactionary' legislation.

His activism took in a considerable number of other causes too. For instance, he dipped his toe into the maelstrom of Middle Eastern politics. Most notably, he was linked to an academic boycott of Israel, pulling out of an appearance at a conference there in 2013. While he did not publicize his decision at the time, a statement from the British Committee for the Universities of Palestine (released with Hawking's approval) described the move as 'his independent decision to respect the boycott, based upon his knowledge of Palestine, and on the unanimous advice of his own academic contacts there'.

He also spoke out against growing economic inequality across the globe. For instance, he responded to a question on a Reddit 'Ask Me Anything' interview about a future in which work is done predominantly by machines: 'If machines produce everything we need, the outcome will depend on how things are distributed. Everyone can enjoy a life of luxurious leisure if the machine-produced wealth is shared, or most people can end up miserably poor if the machine-owners successfully lobby against wealth redistribution. So far, the trend seems to be toward the second option, with technology driving ever-increasing inequality.' It was a depressingly bleak view from a figure at the cutting edge of science.

While he was formally unaffiliated with any political party, his politics were broadly left of centre. Ahead of the 2015 UK general election, he spoke publicly of his intention to vote for the Labour Party, the party to which his mother had formally belonged after her youthful stint with the Communists. He was a stern opponent of the austerity cuts imposed by the Conservative-led governments in power since 2010, particularly in the areas of education and science funding. Indeed, in 2010 he threatened to leave the country if proposed slashes to budgets were carried through. He was also an unapologetic defender of the National Health Service, without which, he said in 2009, he would not be alive today.

Hawking's position on women's issues, however, has been rather more cloudy. He found himself in hot water with at least half of the population in 2005 when he said in an interview with the *Guardian*:

In the past, there was active discrimination against women in science. That has now gone, and although there are residual effects, these are not enough to account for the small numbers of women, particularly in mathematics and physics ... It is generally recognized that women are better than men at languages, personal relations and multi-tasking, but less good at map-reading and spatial awareness. It is therefore not unreasonable to suppose that women might be less good at mathematics and physics. It is not politically correct to say such things and the president of Harvard got in terrible trouble for doing so. But it cannot be denied that there are differences between men and women.

His words seemed curiously out of kilter with his usually socially progressive attitudes, but as we have already seen, it is not the first time he expressed a 'them and us' sentiment regarding women. Furthermore, on other occasions, he came out on the side of gender equality, not least when campaigning in the 1970s for Gonville and Caius to admit female students.

# Work With Your Intuition

'I rely on intuition a great deal.'

STEPHEN HAWKING, SPEAKING ON
*DESERT ISLAND DISCS*, 1992

Given the nature of cosmology – that is to say, attempting to establish laws that govern realms we cannot hope to physically see – it is inevitable that exponents must to some degree rely on their intuition. That is not to say that cosmologists take wild stabs in the dark on the basis of a hunch. Rather, they must process the available data, ruminate on the existing literature and then explore the recesses of their mind in order to formulate hypotheses. Hawking's magnificent – though not infallible – ability to intuit fruitful lines of investigation was a key attribute in his rise to greatness.

While he allowed himself to be open to his intuition, it was only effective because he also demanded it was rooted in a sound rational basis. In the preface to *Black Holes and Baby Universes and Other Essays* (1993), he wrote:

I do not agree with the view that the universe is a mystery, something that one can have intuition about but never fully analyse or comprehend. I feel that this view does not do justice to the scientific revolution that was started almost four hundred years ago by Galileo and carried on by Newton. They showed that at least some areas of the universe do not behave in an arbitrary manner but are governed by precise mathematical laws.

He was also unstinting in stress-testing his ideas and discarded those that broke. 'I try to guess a result, but then I have to prove it,' he said on *Desert Island Discs* in 1992. 'And at this stage, I quite often find that what I had thought of is not true or that something else is the case that I had never thought of.' So it was that even a discarded hypothesis – an intuition that came to nothing – could birth the next great insight or at the very least guide the next phase of research. It is also worth speculating as to whether Hawking's visual way of thinking accentuated his intuitive abilities. By his own admission, he did not 'care much for equations' because he didn't have an 'intuitive feeling' for them. Might it be that his preference for imagining concepts in almost physical terms, rather than as complex equations, gave freer range to that creative imagination in which intuition thrives?

By his own reckoning, his first great intuitive leap – which he variously called a 'Eureka moment' and a 'moment of ecstasy' – came in 1970, as he was putting his young daughter to bed. It was then he experienced a revelation that black holes have entropy, a realization that paved the way for his description of Hawking radiation. It is notable that he hit upon one of his most significant insights away from the confines of a formal academic atmosphere, his creativity blossoming instead when we might suppose his focus was on rather more domestic matters.

Once again, his methods bring to mind Einstein, who in 1920 stated: 'All great achievements of science must start from intuitive knowledge, namely, in axioms, from which deductions are then made … Intuition is the necessary condition for the discovery of such axioms.' In an interview nine years later, he returned to the subject: 'I believe in intuitions and inspirations … I sometimes feel that I am right. I do not know that I am.'

However, while intuition is sometimes understood as being little more than inspired guesswork, both Einstein and Hawking conceived of it as almost the final step in a long process of knowledge accumulation. In Einstein's words: 'Intuition is nothing but the outcome of earlier intellectual experience.'

# Write Like Hawking

'I am pleased a book on science competes with the memoirs of pop stars. Maybe there is some hope for the human race.'

STEPHEN HAWKING, QUOTED IN *STEPHEN HAWKING: A LIFE IN SCIENCE* BY MICHAEL WHITE AND JOHN GRIBBIN, 1992

Hawking's bibliography includes well over 200 books (fiction and non-fiction) and academic papers dating back to 1965. Even taking into account that he was a co-author on a significant number of these works, that is a formidable rate of production. For someone who spent the large part of their career unable to write more than a few words per minute, the achievement takes on another dimension altogether. To say nothing of the fact that one of his books has become among the most celebrated works of popular science in the history of the language.

His first major paper, 'Occurrence of Singularities in Open Universes', published in 1965, signposted the area in which he would make his mark. Within a few years, he was co-publishing with his PhD advisor, Dennis Sciama, and by the time of the appearance in 1973 of *The Large Scale Structure of Space-Time* (written with George Ellis), Hawking was a well-established name in his field. However, needless to say, his output to this stage was all highly technical and virtually impenetrable to those without a prior sophisticated knowledge of cosmology. It was not until the late 1980s – with the appearance of *A Brief History of Time* – that he entered the realm of popular science writing. At a stroke, he introduced millions to a previously abstruse branch of science and changed his own life for ever.

*A Brief History of Time* has its critics. The prose is not always unfailingly elegant and some of the science remains difficult to grasp – though it is hardly his fault that concepts such as imaginary time remain elusive to the layman. It is, popular legend would have us believe, among the most commonly 'unfinished' books by readers. But for all the quibbles, it was a landmark work that broke down an enormously complex subject area – where our universe comes from and where it might be going – into mostly quite digestible bite-size chunks. For all those who never got past the first few pages, many others made it through to the end and were intellectually nourished for their endeavour. It is a fair assumption that it was and will continue to be the inspiration to young, aspiring scientists, some of whom may even yet come to understand the universe better than Hawking himself.

It also paved the way for Hawking to become the pre-eminent science ambassador of his generation. He may have been unable to move or speak of his own volition, but nobody else has come close to making cosmology accessible to the masses. A slew of other popular works followed. *Black Holes and Baby Universes and Other Essays* (dealing with an array of scientific and non-scientific topics) came out in 1993. *The Universe in a Nutshell* helped introduce M-theory into popular discourse in 2001, while *On the Shoulders of Giants* reintroduced a number of classical scientific works to a modern audience a year later. Then, in 2005, he partnered up with Leonard Mlodinow to revisit and update *A Brief History of Time* in the form of *A Briefer History of Time*. The same pairing produced *The Grand Design* in 2010, which incorporated the latest scientific thinking into a reappraisal of the fundamental questions of the universe, including – controversially – the existence (or not) of God.

*My Brief History*, an autobiographical work charting Hawking's gripping personal story and professional adventures, appeared three years later. His last popular science book was 2018's *Brief Answers to the Big Questions*, which his publisher described as 'a selection of [his] most profound, accessible and timely reflections from his personal archive'. There has also been a series of children's fiction books, written with his daughter Lucy, which aim to bring cosmology to a younger audience via the adventures of 'George'. The first title, *George's Secret Key to the Universe*, appeared in 2008, with a further four (*George's Cosmic Treasure Hunt*, *George and the Big Bang*, *George and the Unbreakable Code* and *George and the Blue Moon*) following by 2016. These titles have so far been translated into over a dozen languages. Hawking enjoyed the process of writing for children, an audience he described as 'naturally interested in space and not afraid to ask why'.

All the while, as if to assuage any accusations of 'dumbing down', he continued to write books and papers for more specialist audiences. In 1996, for instance, he wrote *The Nature of Space and Time* (with Roger Penrose) and in 2005 published 'Information Loss in Black Holes' and the long-format *God Created the Integers: The Mathematical Breakthroughs that Changed History*. Meanwhile, his most recent academic papers were the distinctly non-populist 'The Information Paradox for Black Holes', 'Information Preservation and Weather Forecasting for Black Holes' and 'Vector Fields in Holographic Cosmology'. To so successfully write both for the expert and the uninitiated was an achievement to which few scientific writers, if any, have ever come close.

And all of this, lest it be forgotten was achieved in the face of monstrous practical challenges. By the late 1960s his ability to write had already been significantly curtailed by his illness, a situation which only got worse. Then, as his voice weakened over the ensuing years, the process of dictating became more ponderous and draining. When he lost his own voice altogether in the 1980s, he was able to write sentences on his computer. For a while he still had a little use of his thumb but it was not long before his ALS robbed him of even that strength. Then his only means of communication was to spell out words a letter at a time by raising his eyebrow at an appropriate moment as a third party pointed to the relevant letter on a spelling card. One can only imagine the immense frustration this process must have induced. As he stated on his personal website: 'It is pretty difficult to carry on a conversation like that, let alone write a scientific paper.'

Fortunately, new technology intervened and he had a computer fitted to his wheelchair that could be operated by a combination of head, eye and cheek movements. This allowed him not only to write sentences but then transmit them to his voice synthesizer as he desired. At its peak of operation, he was able to impart some fifteen words per minute. A vast improvement on his prior situation, but hardly ideal for an individual whose stock-in-trade was communication. It nevertheless allowed him to write at least one book and several dozen scientific papers. Hawking also contended that his difficulties forced him to be concise and to the point. But as his physical situation continued to deteriorate, his word production rate fell to about a word per minute by 2012. Hawking was subsequently involved in cutting-edge research to develop other means of communication for severely physically handicapped people, including attempts to scan and utilize brainwaves. Nonetheless, such technology remains in its infancy.

## HOW *A BRIEF HISTORY OF TIME* BECAME A PHENOMENON

> 'I knew it was going to be a success when it was translated into Serbo-Croatian.'
>
> STEPHEN HAWKING, QUOTED IN
> *VANITY FAIR*, 2004

The publication of *A Brief History of Time* in 1988 was arguably Hawking's personal 'Big Bang' moment. While he had high hopes of the book doing well, he could not have imagined just how well.

His motivations for writing it were twofold. On the one hand, it was a noble endeavour undertaken in the hope of transmitting the wonder of science to a wide audience. His aim was to show just how far we had come in a few decades towards a fuller understanding of our universal origins. But there was an economic imperative as well. All three of his children were in private education and his personal care (including private nurses) came at a great cost. Meanwhile, his academic salary was comparatively modest and, of course, he faced the spectre of an uncertain future. Therefore, he hoped that a modestly successful popular science book might ease his financial woes and perhaps even put something in the bank for his children. Now, over three decades since its publication, he could console himself that the book achieved all that and more.

As events would play out, the writing of *A Brief History* was tumultuous. His aim was to produce a text accessible to pretty much anyone who might pick up the book. For instance, he wanted to avoid the pages of long equations common to most serious books on cosmology. (Having been warned that each equation he used would potentially halve his readership, the book contains just a single equation – albeit the most famous one in science, $E=mc^2$.) Instead, he wanted to take the pictorial images that he carries in his mind and describe them in plain language, with a few diagrams and simple analogies to ease understanding.

As we have already seen, the process of writing was by no means straightforward even as Hawking first started turning the idea over in his head in 1982. However, the catastrophic bout of ill health he endured in Switzerland in 1985 made the latter stages of writing the book an even tougher task. Yet thanks to his unstinting energy and determination, he saw the project through to its conclusion.

The first publisher to offer on the book was Hawking's regular publisher, Cambridge University Press. It offered what was a very healthy advance of £10,000, but this did not correspond with the ambitions its author had for the title. Hawking wanted this to be the sort of book you see piled up in airport bookshops – a real mass-market smash. He therefore decided to engage the services of a New York super-agent by the name of Al Zuckerman. But even Zuckerman suspected his client was overreaching himself.

The year was 1984 and Zuckerman was armed with Hawking's first draft. It was not long before there was a new offer on the table, from the well-respected Norton publishing house. Zuckerman urged Hawking to take the deal but Hawking was determined to hold out. His intuition was soon proved well founded. Bantam, a firm with mass-market reach but little in the way of scientific heritage, came in with a bid of US$250,000 upfront for US rights alone. Hawking did not need to think twice. When he met with his editor-to-be, Peter Guzzardi, in 1985, Guzzardi could not stop himself effusively singing the praises of his newest author. Hawking, though – speaking through a translator – cut him short with the line: 'Where's the contract?'

Guzzardi took an active role in reconfiguring the draft text and the presses were finally ready to roll in 1988. It was also he who suggested adapting Hawking's original title (*From the Big Bang to Black Holes: A Short History of Time*) to *A Brief History of Time: From the Big Bang to Black Holes*. Fortunately, editor and author also decided to reconsider the decision to cut the line that has become the most famous in the book, in which Hawking talks of 'knowing the mind of God'.

An introduction by another living icon of science, Carl Sagan, also helped to raise readers' expectations.

The first inkling that the book was doing something special came when a knowledgeable reviewer from the science journal *Nature* phoned Bantam. He warned that the pre-publication copy he had been sent contained several typographical errors and mislabelled diagrams. The publisher decided to reprint and so recalled the first print run. However, a significant number of copies had already been snapped up. *A Brief History* was clearly hot property. Sales quickly gained momentum and the book went on to break countless records for a science title. In the USA, for example, it was on the *New York Times* bestseller list for 147 weeks and in the UK it stayed in the *Times'* bestsellers for a record-breaking 237 weeks. It has been translated into over forty languages (over forty different ways to be mystified by imaginary time!) and has sold somewhere between 10 and 25 million copies. As Nathan Myhrvold was quoted in *Vanity Fair* in 2004: 'It outsold Madonna's book *Sex*, and by a huge margin, and who would have predicted that?'

# Enjoy Your Celebrity

'Being well known and easily recognizable had its pluses and minuses … But the minuses are more than outweighed by the pluses.'

STEPHEN HAWKING, *MY BRIEF HISTORY*, 2013

Even as the *Brief History* bandwagon began to roll, Hawking surely had no concept that he was about to be elevated to the status of bona fide superstar. He joined an exclusive club of scientists whose name, face (and of course, voice) are instantly recognizable on every continent around the world.

Fame tends to be a double-edged sword, but Hawking adapted to it spectacularly well. He enjoyed the trappings of success (and its accompanying wealth) and why should he not? But he also used his celebrity to promote a great many causes in which he firmly believed. Most importantly of all, he continued to take science to places that only the power of celebrity can reach. Nonetheless, he found himself subject to the slings and arrows that any other mega-celebrity – from Beyoncé or Ronaldo to Barack Obama – could expect to have fired at them.

Hawking's critics tended to focus on two particular and related criticisms in relation to his fame. The first – and one that centres on the 'elephant in the room'

– is that Hawking would not have been as famous as he was if he were not disabled. It is a question that Hawking himself pondered, not least in a 2014 PBS documentary: 'Sometimes I wonder if I'm as famous for my wheel-chair and disabilities as I am for my discoveries.'

There is no doubt that his disability played a role in maintaining the public's interest in him. That an individual could be at once so cruelly disabled and so brilliantly able is of endless fascination to the world at large. This 'human interest' aspect of his life served to provide a way into his science for many people who might not otherwise give a moment's thought to the latest thinking among cosmologists. Hawking should not have had to answer for his disability or justify the level of fame he attracted as a result of it. However, there are those who suggested that because he garnered such monumental attention, other scientists who are doing work of equal or greater magnitude went unrecognized.

More problematically, some of his critics suggested that his name was so powerful that his particular theoretical positions were received uncritically, while competing theories were brushed aside. Again, it was hardly his fault that not everyone got an equal share of the spotlight, any more than it is Lionel Messi's 'fault' that he is lavished with more praise than teammates who may on occasion have a better

## LIFE IN THE FAST LANE

Despite having to deal with occasional barbs, Hawking nonetheless seemed to embrace life in the fame fast lane. In truth, there would be little point in attempting to shy away from it, unless he was prepared to live the life of a hermit. As he archly told Israeli television in 2007: 'The downside of my celebrity is that I cannot go anywhere in the world without being recognized. It is not enough for me to wear dark sunglasses and a wig; the wheelchair gives me away.' So it is that he was happy to mix with fellow A-listers (Richard Branson and Daniel Craig were among the guests at his seventieth birthday party) and to give lectures in venues more used to hosting pop stars. He sold out London's Royal Albert Hall, 'gigged' in the White House and in 2014 headlined Starmus – a 'cosmos and music fest' held in the Canary Islands – opposite guitar legend and keen amateur astronomer, Brian May. Back in 1990 Hawking displayed an admirable streak of rock star-esque attitude of his own when he blew off a planned press conference ahead of a lecture in the British seaside town of Brighton in order to see a Status Quo concert.

game than him. However, it should perhaps have put additional pressure on him to self-edit his public communications. Being something of a showman, that was not always the case. As his old friend, Martin Rees, put it: 'A downside of celebrity is that his comments attract exaggerated attention even when he speaks about topics in which he has no special expertise.'

Celebrity was an intrinsic and self-perpetuating part of the Hawking phenomenon, but it was a distant second to his work as a scientist. Returning to the words of Martin Rees: 'His fame should not overshadow his scientific contributions, because even though most scientists are not as famous as he is, he has undoubtedly done more than anyone else since Einstein to improve our knowledge of gravity.'

## HAWKING THE TV GUEST STAR

'I was happy to show that science can also have street cred.'

STEPHEN HAWKING, ON APPEARING IN
*THE SIMPSONS*, 2004

If you were asked to describe the archetypal television personality, you probably would not immediately go for a severely physically disabled theoretical physicist who spoke through a curiously accented voice synthesizer.

But then, very little in the world of Hawking was typical. Despite the seemingly unnatural fit with TV, he carved out a significant small-screen career that lasted well into three decades. Furthermore, it turned out that he had range too – his appearances encompassed everything from documentaries and talk shows to drama and comedy.

For all that he didn't meet the identikit image of a TV star, he nonetheless had several things going for him. Firstly, the success of *A Brief History* made him a name (or even *the* name) that the public associate with science, intelligence and indeed 'genius'. In addition, he was immediately recognizable – despite Hawking's witticisms, nobody mistook him for someone else. This was the case both visually and aurally. His synthesized voice became an integral part of Hawking the public figure. So much so that he refused all offers to give his voice a more naturalistic tone using upgraded software.

In terms of his television work, the projects of most enduring value are the various landmark documentary series to which he was attached. In 1997 he fronted PBS's *Stephen Hawking's Universe*, in which he provided a broad survey of developments in astronomy and cosmology. Then in 2008 there was a short series for Channel 4 called *Stephen Hawking: Master of the Universe* – a mixture of biography and hard science that brought in healthy UK viewing figures of close to 2 million, a figure undreamed of for most science documentaries. *Into the Universe with Stephen Hawking* for the Discovery Channel followed in 2010, in which Hawking guided viewers across the universe with the aid of cutting-edge, computer-generated imagery. *Brave New World with Stephen Hawking*, again for Channel 4, came out the following year, looking over five episodes at how scientists are attempting to make the next leaps forward in fields including technology, health and the environment. Keeping up an unrelenting pace, he next presented *Stephen Hawking's Grand Design* (for the Discovery Channel), a series in part based on his book of the same name and addressing many of the big questions that dominated his research career. Meanwhile, in 2007 he found the time to provide introductions to a series of stand-alone dramas in *Masters of Science Fiction*.

His screen fame was further cemented by major biographical documentaries (including *A Brief History of Time* in 1991 and *Hawking* in 2013) and by fictional representations of his life. The 2004 BBC drama *Hawking* was much praised at the time and saw Benedict Cumberbatch (pre-*Sherlock* superstardom) in the title role. Then the Oscar-nominated *The Theory of Everything* was released in 2014, with Eddie Redmayne playing the lead. Hawking's fame was instantly guaranteed for another generation at least.

Until *The Theory of Everything*, however, Hawking was arguably best known (on screen at least) for his cameo appearances in the long-running US animated comedy, *The Simpsons*. First showing up in the episode 'They Saved Lisa's Brain', he was variously portrayed as a plagiarist (stealing Homer Simpson's idea for a doughnut-shaped universe), a wise restaurant owner, a confused maze-navigator, a rapper, and as the owner of a wheelchair replete with a spring-loaded boxing glove and helicopter blades. If proof were needed that Hawking enjoyed a joke, this was it. It also made him realize how fickle fame can be, as he wryly acknowledged '… people think I'm a *Simpsons* character'.

There followed further cameos in another animated series from *The Simpson's* creators, *Futurama*. Then the smash-hit US comedy *The Big Bang Theory* (based around the lives of a group of scientists studying and working at Caltech) gave him the opportunity to show off his skills in a live-action show, not least when he described a calculating error by the show's self-professed resident genius, Sheldon Cooper, as 'quite the boner'. Hawking's cult comedy status was reinforced in 2014 when he appeared with the reunited Monty Python gang for a sold-out show in London, using his wheelchair to run over fellow scientist and broadcaster, Brian Cox.

Hawking also memorably lived out the classic science

fiction fan's dream of appearing on *Star Trek* when he featured in a 1993 episode of *The Next Generation* – playing poker against Data, Isaac Newton and Albert Einstein, and so became the first person ever to play himself on the show. Meanwhile, in 2015 he revealed to *Wired* what he considered would be his perfect film role: 'My ideal role would be a baddie in a James Bond film. I think the wheelchair and the computer voice would fit the part.'

# Looking to the Future: Is Time Travel Possible?

'I'm obsessed by time. If I had a time machine I'd visit Marilyn Monroe in her prime or drop in on Galileo as he turned his telescope to the heavens. Perhaps I'd even travel to the end of the universe to find out how our whole cosmic story ends.'

STEPHEN HAWKING, *DAILY MAIL*, 2010

Did Hawking believe in time travel? The answer is a decidedly equivocal 'Yes. And no.' He stated his nervousness about talking about time travel in public, given the propensity for the subject to be treated as somehow 'unserious' and for its advocates to be labelled as cranks. As he noted, even if it turns out that time travel is not possible, it is important that we should understand the reasons why. One of the ways to get around this problem, Hawking suggested, is to rebrand time travel as 'particle histories that are closed'!

So where did he stand on the question? On the one hand, he didn't have much time for the theory that humans can nip backwards and forwards at will across the centuries. On the other hand, he did believe that the nature of the universe allows for theoretical forward jumps through time.

To deal with the 'intrepid time traveller' idea first, he was highly sceptical that time travel as envisaged in your average sci-fi novel will ever be possible. If it were, he contended, where are the hordes of tourists from the future? Consider his words to delegates at the Seattle Science Festival in 2012: 'I have experimental evidence that time travel is not possible. I gave a party for time travellers, but I didn't send out the invitations until after the party. I sat there a long time, but no one came.'

If his justification for his assertion was somewhat wryly put, it is not without validity. Furthermore, Hawking suspected that natural laws prohibited backwards time travel in order to prevent the occurrence of what are known as paradoxes. To take the classic example of a paradox, what would happen if you could travel back in time and kill your grandmother in her youth? If you kill your grandmother, she will not have the daughter who is your mother and so nor will you be born. But if you are not born, how could you have been alive to time travel in the first place and rewrite history? An endlessly mind-boggling conundrum. (In Hawking's version, a mad scientist goes backwards in time and kills his younger self before he has cracked the mysteries of time travel.)

To deal with the prospect of paradoxes, Hawking devised the Chronology Protection Conjecture (see page 165). However, he was less dismissive of the notion that it may be – just may be – possible to travel forwards in time at an accelerated rate. To explain how, we must look to Einstein's general theory of relativity. Where Sir Isaac Newton conceived of time in absolute terms, Einstein argued it was a rather more pliable concept. In his general theory, he outlined the existence of space-time which is curved and distorted by all the matter and energy that exists within the universe. While 'time' at the local level moves in a forward direction, in Einstein's vision of the universe it could potentially become so distorted that it curves back in on itself.

Much of modern science accepts the existence of hypothetical tunnels – commonly known as wormholes – that link different points in space-time. Wormholes effectively connect two different places and times so that were it possible to travel down one, you could take a shortcut to a different moment in history. The only trouble is, as Hawking acknowledged, wormholes are thought to exist only at the sub-atomic level in what is known as the quantum foam. To access a wormhole, you'd need to measure no more than a billion-trillion-trillionths of a centimetre. Even Alice couldn't manage that in Wonderland, and she had a magical shrinking potion.

Nonetheless, there is a body of scientific thought which postulates that future technologies may make it possible to enlarge a wormhole sufficiently to allow a human through. Hawking, however, doubted it. He suspected that such a process of enlargement would create radiation feedback that would grow so quickly that the wormhole would be destroyed before anyone could enter it. But he did not give up on time travel altogether.

As with so much of Hawking's work, he believed one possible solution to the problem lay with black holes. As Einstein's general theory elaborated, time travels at different speeds in different parts of the universe. We know this to be true from super-precise clocks that have been fitted to satellites that comprise our GPS system high above in Earth's orbit. These clocks lose a tiny fraction of a second each day because time in space runs faster than it does here on Earth. This is because heavy objects (such as our planet) act as a drag on time.

If you are looking for really heavy objects, look no further than black holes, or better still, a supermassive black hole. We have one in the middle of our galaxy, about 26,000 light years away. It contains mass equivalent to four million suns condensed to a single point. And it makes time go much slower than we are used to. If it were possible to send humans to fly round this black hole, those on board would experience time at half the speed of those they left behind. If the crew stayed in orbit for a year, they would return to a planet which had gone through two years.

However, travelling that close to a black hole is a risky business, and the time-leap achieved would be, relatively speaking, quite small. Instead Hawking suggested our best hope of significant forward time travel is to invent a machine that travels at currently unimaginable speeds. Physics allows nothing to exceed the speed of light (about 186,000 miles per second). This means that as an object approaches that speed, time starts to go slower to ensure it remains within the speed limit. So, for instance, if a spaceship was travelling at 99.9999% of the speed of light and someone on board that ship was running at sufficient velocity that they might surpass the speed of light, time on board will effectively put them into slow motion. Travel fast enough, and what might be just a few hours on the ship could equate to years on Earth.

To achieve the required speed will take some doing. The US rocket *Apollo 10* is the fastest manned vehicle ever produced, and we would need something about 2000 times faster. It would also need a good six years of travelling at full throttle to build sufficient speed. Yet it is, according to Hawking, our most promising means of time travelling and would also open up the universe to us in hitherto unimagined ways.

## THE CHRONOLOGY PROTECTION CONJECTURE

'It seems that there is a Chronology Protection Agency which prevents the appearance of closed time-like curves and so makes the universe safe for historians.'

STEPHEN HAWKING, *PHYSICAL REVIEW*, 1992

For people of a certain age, it was Marty McFly and the *Back to the Future* movies that introduced them to the concept of the time-traveller's paradox. In the first movie of that terrific trilogy, McFly finds himself in the past before his own birth and must orchestrate his prospective parents' courtship in order to guarantee his own conception in the future. It is thus a high-concept movie in all respects.

Over the years, a few unsatisfactory semi-solutions to the paradox conundrum have been proposed by scientists and science-fiction writers alike. Might it be possible, for instance, to create a sort of virtual reality machine that replicates the past without actually impinging upon it? Or could we visit what appears to be our past but in a parallel universe, so as to leave our actual past undisturbed? In each case, we visit a version of the past but not our own, real past.

Legend has it that when Carl Sagan was writing his time-travel novel, *Contact*, he got in touch with Kip Thorne in the hope of ironing out some time-travel plot holes. It got Thorne thinking about the possibility of time travel via wormholes, which he elucidated upon in a paper published in 1988. The media immediately dubbed him 'The Man Who Invented Time Travel' even though Thorne himself was among the loudest voices to argue that practical time travel was almost certainly impossible since quantum gravity would likely result in the destruction of a time machine as soon as it was put into action. This conclusion in turn eventually led on to Hawking's own observation contained in the quotation at the start of this section.

While backed by solid science, Hawking's Chronology

Protection Agency is resolutely conjecture rather than law. Quite how – and when – we will ever be able to test the hypothesis is anyone's guess.

# Beware, the End is Nigh (But Probably Not That Nigh)

'The development of full artificial intelligence
could spell the end of the human race.'

STEPHEN HAWKING, INTERVIEW WITH
THE BBC, 2014

Given his unique status as the go-to sage of cosmology, it is inevitable that Hawking was regularly asked what he considered the greatest threats to our future. Unfortunately, the more sensationalist elements of the media have a tendency to spin his comments on the subject into wild predictions of imminent doom for the world. Hawking, needless to say, was rather more circumspect about the threats. Nonetheless, he gave some stark warnings as to the potential dangers of letting our technological advances get the better of us.

We have already looked at Hawking's dislike for nuclear weapons, but it is one of several potential paths to Armageddon that he envisaged. For example, in common with the vast majority of serious scientists, he was fearful that climate change – in large part driven by increased greenhouse gas emissions as our modern world makes ever greater energy demands – could render our current way of life unrecognizable. In 2007, he spoke at a press conference for the Bulletin of the Atomic Scientists on the occasion of the Doomsday

Clock (a symbolic clock that suggests a countdown to potential global disaster) being moved forward by two minutes to show five minutes to midnight (the hour of disaster). He said:

As scientists, we understand the dangers of nuclear weapons and their devastating effects, and we are learning how human activities and technologies are affecting climate systems in ways that may forever change life on Earth. As citizens of the world, we have a duty to alert the public to the unnecessary risks that we live with every day, and to the perils we foresee if governments and societies do not take action now to render nuclear weapons obsolete and to prevent further climate change ... There's a realization that we are changing our climate for the worse. That would have catastrophic effects. Although the threat is not as dire as that of nuclear weapons right now, in the long term we are looking at a serious threat.

He also registered his concern about the dangers of unregulated bioengineering, a field of research that has made exponential advances in recent years. Indeed, back in 2001 he suggested in an interview with the *Daily Telegraph* that unfettered biological research worried him more than the threat of nuclear annihilation. 'Nuclear weapons need large facilities,' he said, 'but genetic engineering can be done in a small lab. You can't regulate every lab in the world. The danger is that either by accident or design, we create a virus that destroys us.'

But it is the threat posed by artificial intelligence (AI) to which he returned with most regularity. Given the sometimes rather woolly definitions of what constitutes AI, it is a subject area fraught with potential misinterpretation. So what do we mean by the term? AI refers to the intelligence exhibited by a machine or software. Those working in the field thus aim to create increasingly 'intelligent' machines – that is, machines with the ability to learn and adapt their behaviour to suit particular circumstances and so guarantee the most effective outcomes. This is, in a certain light, the stuff of sci-fi nightmares, conjuring up images of out-of-control super-robots overhauling their human former overlords. One might refer to the 2004 movie, *I, Robot*, for a classic example of AI paranoia. Hawking's concerns, however, were rather more subtle than terror of renegade 'bots. Consider, for instance, his response to a question posed in his 2015 Reddit 'Ask Me Anything' (AMA) interview:

The real risk with AI isn't malice but competence. A super-intelligent AI will be extremely good at accomplishing its goals, and if those goals aren't aligned with ours, we're in trouble. You're probably not an evil ant-hater who steps on ants out of malice, but if you're in charge of a hydroelectric green energy project and there's an anthill in the region to be flooded, too bad for the ants. Let's not place humanity in the position of those ants. Please encourage your students to think not only about how to create AI, but also about how to ensure its beneficial use.

Hawking was by no means anti-AI per se. Instead, he was nervous that technological developments are greatly outpacing our philosophical considerations. As he observed in a speech at the 1994 Macworld Expo in Boston, 'I think computer viruses should count as life … I think it says something about human nature that the only form of life we have created so far is purely destructive. We've created life in our own image.' So how much autonomy should humans cede to machines in the interests of increased efficiency? Some of the biggest ethical questions we face surround the potential development of 'autonomous weapons' (or 'killer robots' as they are sometimes known). Several steps on from the armed drones that are already in wide use, autonomous weapons could theoretically select and engage targets without any human involvement. War, in other words, would be conducted by machines – slaughter outsourced so as to make it all the more difficult to trace responsibility for specific operational decisions back to particular individuals.

While there seems to be a broad consensus among governments at the moment to steer clear of autonomous weapons, it is impossible to state that this will always be the case. In the event that one rogue nation or militant group comes into possession of such technology, it is hard to envisage that other agencies won't feel they have to keep apace. In July 2015 Hawking was among the most prominent of over a thousand experts to sign a letter presented to the Twenty-Fourth International Joint Conference on Artificial Intelligence, which warned against an AI arms race and called for an outright ban on autonomous weapons. Other notable signatories included Apple co-founder Steve Wozniak, Tesla Motors CEO Elon Musk, and CEO of Google DeepMind, Demis Hassabis.

The problem, as Hawking saw it, is that we just cannot predict what twists and turns may occur along the AI path into the future. Specifically, we cannot know the full implications of creating a machine that is cleverer than the people who made it. As he told the BBC in 2014:

> We cannot quite know what will happen if a machine exceeds our own intelligence, so we can't know if we'll be infinitely helped by it, or ignored by it and sidelined, or conceivably destroyed by it.

Contrary to those headline-writers who opted to

depict him as a harbinger of certain AI-induced doom, Hawking simply urged caution. Rather than blindly striving towards the creation of super-intelligent machines, we should devote time now to considering how these machines will materially affect our lives – for better and worse – and figure out how we can ensure they serve only to improve our collective existence. Responding to another question in his 2015 AMA interview, he said:

> We should shift the goal of AI from creating pure undirected artificial intelligence to creating beneficial intelligence. It might take decades to figure out how to do this, so let's start researching this today rather than the night before the first strong AI is switched on.

# Are Advanced Civilizations Fated to a Short Existence?

'It is not clear that intelligence has any
long-term survival value.'

STEPHEN HAWKING, 'LIFE IN THE
UNIVERSE' LECTURE, 1996

Hawking exhorts us to consider the dangers of what he saw as the leading existential threats in order that humanity may endure. However, he also raised a related and disturbing possibility – that advanced, intelligent life forms (such as humans) are inherently prone towards self-destruction. The more technological advancements we achieve, the greater the risk to our continuing existence.

Hawking was not always kind about our species as a whole ('Primitive life is very common and intelligent life is fairly rare. Some would say it has yet to occur on Earth,' he jested in a 2008 lecture at George Washington University) but he did recognize our – as far as we know – unique status in terms of intelligence. In 1988, for instance, *Der Spiegel* reported this rather backhanded compliment from him: 'We are just an advanced breed of monkeys on a minor planet of a very average star. But we can understand the universe. That makes us something very special.'

It is our apparent uniqueness in this respect that

serves as cause for concern. After all, if it was truly beneficial to develop intelligence, wouldn't there be more intelligent species around? Speaking at the Seattle Science Festival in 2012, he put it in these terms:

> We don't know the probability that a planet develops life. If it is very low, we may be the only intelligent life in the galaxy. Another frightening possibility is intelligent life is not only common, but that it destroys itself when it reaches a stage of advanced technology. Evidence that intelligent life is very short-lived is that we don't seem to have been visited by extraterrestrials. I'm discounting claims that UFOs contain aliens. Why would they appear only to cranks and weirdos? … Further evidence that there isn't any intelligent life within a few hundred light years comes from the fact that SETI, the Search for Extraterrestrial Life, hasn't picked up their television quiz shows.

Hawking long argued that in order to safeguard the long-term future of the species, humans must prepare to spread beyond Earth – and to do so sooner rather than later. Back in 2001, he told the *Daily Telegraph* that the planet is simply too vulnerable to guarantee our ongoing survival, since life is subject to the possibility of too many accidents. His prediction then was that the human race needs to travel beyond Earth's frontiers within a millennium. In 2010, in an interview with the Big Think website, he was even more pessimistic, suggesting we have a two-hundred-year window in which to colonize space if we are not to die out. By 2015, when he took part in a televised interview with physics graduate and comedian Dara Ó Briain, even his hopes that we might achieve this seemed to have subsided. Back in 2001 he had declared himself an optimist and proclaimed, 'We will reach out to the stars.' But now he asserted, 'The present breed of humans won't reach the stars,' justifying his position on the grounds that the distances involved are too great and the likely radiation exposure too severe. Our only hope, he suggested, is if we are able to genetically engineer humans or send machines to undertake a colonization for us.

He painted a bleak picture but we should perhaps take his words with a small pinch of salt. This was, after all, a high-profile interview and he was aware that such bold utterances were sure to elicit wider attention for himself and his interests. Furthermore, Hawking was no more able to predict the future than anybody else. His analysis of humanity's future can only ever be speculative. But by espousing 'worst-case scenarios', he posited a vital debate – how might we mould our future as a species? – firmly into the public sphere. If Hawking could sometimes be accused of scaremongering, it was with a purpose in mind. Consider, for instance, an open question he posted on Yahoo Questions in 2006: 'In a world that is in chaos politically, socially and environmentally, how can the human race sustain another hundred years?' 'I don't know the answer,' he later clarified. 'That is why I asked the question, to get people to think about it, and to be aware of the dangers we now face.'

His question garnered some 25,000 responses within a few weeks. His favourite of these indicated he may have had rather higher hopes for us than he sometimes admitted to. Proffered by a poster calling themselves Semi-Mad Scientist, it read: 'Without the belief that we will continue to grow and overcome the pains of social chaos as we mature as a species, we might as well not have any faith at all. I'm not talking religion … but simply the same belief that we will survive just as much as the sun will rise the next day.'

## IS THERE ANYBODY OUT THERE?

> 'We only have to look at ourselves to see how intelligent life might develop into something we wouldn't want to meet.'
>
> STEPHEN HAWKING, *INTO THE UNIVERSE WITH STEPHEN HAWKING*, 2010

Hawking's argument that space colonization is desirable naturally leads on to another question: might we expect to bump into anyone else? In broad terms, he believed that the balance of probability favours the existence of extraterrestrial life but not necessarily intelligent extraterrestrials.

Given that we know there are hundreds of billions of stars spread across hundreds of billions of galaxies, it seems unlikely – not to mention vain to assume – that it is only our little planet that supports life. But by the same token, given that we are the most advanced species of the millions that exist on Earth and that about half of the planet's biomass consists of bacteria-like organisms, it is reasonable to assume that the large proportion of all life is primitive. Indeed, if his thesis that intelligence is not only not necessary to survival but may actively imperil it, then we should assume most successful life forms are non-advanced.

However, we of course lack empirical evidence for such an assertion so Hawking refused to rule out the existence of other advanced life forms elsewhere in the universe. And given that we could not possibly second-guess their capabilities and intentions, he urged extreme caution in attempting to make contact with them. If intelligent life does exists, what is to say that it is not far more highly developed than we are and desirous of turning our species into its slaves, just as we have exploited other species, and even races, on our own planet? In his documentary series *Into the Universe*, he went as far as to suggest that if aliens were to visit us, the outcome would be 'much as when Columbus landed in America, which didn't turn out well for the Native Americans'. This echoed his words several years earlier on the National Geographic programme, *Naked Science: Alien Contact*, 'I think it would be a disaster. The extraterrestrials would probably be far in advance of us. The history of advanced races meeting more primitive people on this planet is not very happy, and they were the same species. I think we should keep our heads low.'

So we are left with the somewhat curious paradox (hardly an unusual phenomenon in Hawking's story) where, on the one hand, he exhorted humanity to go out into the universe but, on the other hand, suggested we keep a low profile on the cosmological stage. Then, in 2015, he threw an additional curve ball as he publicly committed himself to a £100-million-project – led by two Hawking allies of long-standing, Martin Rees and Yuri Milner – dedicated to seeking out other intelligent life in the universe. Over ten years, the project, called Breakthrough Listen, aims to survey the million closest stars to Earth in search of life, as well as to listen out for messages from the hundred nearest galaxies. At its launch, Rees said, 'It's a huge gamble of course, but the pay-off would be colossal.'

Speaking at the same event, Hawking said:

To understand the universe you must know about atoms, about the forces that bind them, the contours of space and time, the birth and death of stars, the dance of galaxies, the secrets of black holes. But that is not enough. These ideas cannot explain everything. They can explain the light of stars but not the lights that shine from planet Earth. To understand these lights you must know about life, about minds. We believe that life arose spontaneously on Earth. So in an infinite universe there must be other occurrences of life. Somewhere in the universe intelligent life may be watching the lights of ours aware of what they mean. Either way there is no better question. It is time to commit to finding the answer to life beyond Earth. We are life, we are intelligent, we must know.

Perhaps Hawking concluded that we are ultimately best served by getting to know what's out there – whether friend or foe – before they know everything about us.

# Never Stop Challenging Yourself

'I have had a full and satisfying life.'

STEPHEN HAWKING, *MY BRIEF HISTORY*, 2013

As Hawking reached his eighth decade, most of which had been spent battling appalling physical symptoms, he inevitably slowed down a little. Health concerns meant that his diary of public appearances was subject to change at short notice. Each passing day represented a significant achievement as he battled the ALS that doctors predicted would kill him fifty or more years previously. But arguably most remarkable of all was the fact that Hawking remained a part of public life. While even those of his age who boast rude health might think the time has come to take a back seat, he kept on going – as an academic and as a public ambassador for science.

Hawking always possessed a spirit of adventure and a lust for life. For instance, while he was a student he somehow found himself caught up in a scheme to smuggle Russian-language Bibles into the Soviet Union. When the authorities got wind of the operation, the young Hawking was held for several hours before being released. Looking back on the experience, his overriding memory seemed to be of a sense of exhilaration. The desire to 'experience'

– even when it might not be comfortable – stayed with him throughout his life.

Despite the obstacles thrown up by his disability, he has travelled on every continent on the planet bar Australasia. Even as he entered his senior years, he continued to be a regular visitor to research institutions in Europe and North America. Back in 1997 he even undertook a trip to Antarctica, an adventure he subsequently nominated as his 'favourite memory'. The purpose of the visit was to attend a high-level conference of theoretical physicists from around the globe. Hawking was flown by the Chilean Air Force to their base on Isla Rey Jorge in the Antarctic Peninsula. Once there, he was ferried around by snowmobile – a mode of transport he found far more agreeable than his wheelchair, which had been modified for the occasion with snow chains. It is easy to see how the escapade appealed to Hawking's particular sense of fun.

The trip to Antarctica and the gravity-free flight mentioned overleaf are just two adventures in a lifetime punctuated by seriously challenging tests – whether of intellect, stamina, bravery or determination. Hawking was rarely found wanting. Consider, for instance, his insistence on visiting a state-of-the-art neutrino laboratory located about a mile and a half underground near Sudbury, Ontario, in Canada. In order to access the facility, he travelled in a specially commissioned elevator car that plunges deep into the earth. The observatory's enthusiasm to host the iconic scientist was matched only by his enthusiasm to see their groundbreaking work at first-hand.

---

### ONE STEP CLOSER TO SPACE?

In 2007, Hawking again confounded medical opinion by taking a zero-gravity flight in a specially modified Boeing 727. The experience, which apes the sensation of flying in space, can cause unfortunate side-effects in passengers, resulting in the aircraft being known colloquially as the 'Vomit Comet'. However, the images of Hawking floating freely, liberated from his wheelchair, linger long in the mind. Speaking after the two-hour flight, which involved him experiencing weightlessness in eight twenty-five second bursts (instead of the intended maximum of three, such was Hawking's enthusiasm), he said, 'It was amazing. The zero-G part was wonderful and the higher-G part was no problem. I could have gone on and on. Space, here I come!' Nor were these merely empty words – Hawking had a reservation for a sub-orbital flight with Richard Branson's Virgin Galactic project.

---

As the man with the golden touch when it comes to promoting public engagement with science, Hawking remained in the public eye in his later years, making high-profile television appearances, subjecting himself to open-access online interrogations and conducting countless traditional media interviews. In August 2015 he hit the headlines when he gave a lecture at the Hawking Radiation Conference in Stockholm, Sweden, in which he asserted that black holes 'are not the eternal prisons they were once thought. Things can get out of a black hole both on the outside and possibly come out in another universe.'

In this lecture, Hawking set the scene for his forth-coming paper – developed in partnership with Malcolm Perry and Andrew Strominger – which he suggested conclusively resolved the information paradox. While excitable media claims that he had 'discovered a way to escape from a black hole' were some way wide of the mark, and some fellow experts suggested that elements of what he said were not brand new, it nonetheless highlighted Hawking's unrivalled ability to remain on the cutting edge and to excite popular interest in even the most esoteric subject areas. His unique position among scientists was again emphasized in 2016 when he became the 'go-to guy' for comment following the announcement that gravitational waves originally posited by Einstein a century earlier had been detected for the first time.

Hawking's career took off in the 1960s when he was labouring under the impression that his life would likely be a short one. Over the ensuing decades, his fight against ALS came at no small cost to himself, physically and emotionally. But he consistently embraced the extra time he had been given, exploiting opportunities as they came his way and eternally aiming at the next goal.

# Hawking
# and God

'There is a fundamental difference between religion, which is based on authority, and science, which is based on observation and reason. Science will win, because it works.'

STEPHEN HAWKING, IN AN INTERVIEW
WITH DIANE SAWYER, 2010

Without doubt, the single most famous line that Hawking ever wrote is the one that concludes *A Brief History of Time*. Referring to the 'theory of everything', he said, 'If we find the answer to that, it would be the ultimate triumph of human reason – for then we would know the mind of God.' Yet those memorable and evocative words contain a self-contradiction. For in Hawking's conception of the universe, he saw no inherent need for the existence of God.

This is necessarily a controversial position, which has upset the faithful as much as it has consoled non-believers over the years. In fact, Hawking hardened his stance with the passage of time. Where once he argued merely that the universe could exist independent of a godhead, he did not specifically deny the existence of a deity. More recently, however, he overtly stated his personal atheism.

For comparison, let us consider some of the remarks he made about religion in his earlier career. For example, in *A Brief History of Time*, he wrote, 'The whole history of science has been the gradual realization that events do not happen in an arbitrary manner, but that they reflect a certain underlying order, which may or may not be divinely inspired.' In an interview with *Der Spiegel* in 1988, he elaborated on the theme:

What I have done is to show that it is possible for the way the universe began to be determined by the laws of science. In that case, it would not be necessary to appeal to God to decide how the universe began. This doesn't prove that there is no God, only that God is not necessary.

As recently as 2007, he maintained this relatively moderate position in which he casts doubt on the notion of an omnipotent, creator God, without ruling out the notion of God altogether. As the *New Statesman* quoted him: 'I'm not religious in the normal sense. I believe the universe is governed by the laws of science. The laws may have been decreed by God, but God does not intervene to break the laws.'

Such comments conceivably allowed Hawking to adopt a position on religion reminiscent of that taken by Einstein. He espoused a pantheistic doctrine in which the universe is viewed as a physical manifestation of the divine. Describing his belief system, Einstein once wrote, 'I cannot conceive of a personal God who would directly influence the actions of individuals or would

sit in judgement on creatures of his own creation ...
My religiosity consists of a humble admiration of the
infinitely superior spirit that reveals itself in the little
that we can comprehend about the knowable world.'
Tellingly, he also condemned those who aggressively
championed atheism, describing them as 'creatures
who can't hear the music of the spheres'.

But as Hawking aged, he significantly diverged from
such an Einstein-esque line. For instance, while hosting
an episode entitled 'Did God Create the Universe?' in
the Discovery Channel's 2011 *Curiosity* series, Hawking
argued, 'We are each free to believe what we want and
it is my view that the simplest explanation is there is no
God. No one created the universe and no one directs
our fate. This leads me to a profound realization. There
is probably no heaven, and no afterlife either.'

Where Einstein could never entirely dismiss the
idea of a higher 'divine' order at play in the universe,
Hawking became convinced by the weight of scientific
evidence that God is not apparent in the workings of
the universe. As he put it in *The Grand Design*:

Because there is a law such as gravity, the universe
can and will create itself from nothing. Spontaneous
creation is the reason there is something rather than
nothing, why the universe exists, why we exist. It is
not necessary to invoke God to light the blue touch
paper and set the universe going.

God, then, becomes a construct of the human imagination that, as Hawking saw it, goes against logic, serving instead to provide solace for those seeking order in their personally disordered worlds. As he put it to the BBC in 1990: 'We are such insignificant creatures on a minor planet of a very average star in the outer suburbs of one of a hundred thousand million galaxies. So it is difficult to believe in a God that should care about us or even notice our existence.'

Having faced the prospect of death from an early age, Hawking had a long time to contemplate the likelihood of an afterlife. His conclusion, as he told Dara Ó Briain in 2015, is that it is nothing but 'a fairy story for people afraid of the dark'. This reiterated the sentiments he expressed in an interview with the *Guardian* four years earlier, when he said:

> I have lived with the prospect of an early death for the last forty-nine years. I'm not afraid of death, but I'm in no hurry to die. I have so much I want to do first … I regard the brain as a computer which will stop working when its components fail. There is no heaven or afterlife for broken-down computers …

Hawking's scepticism about religion came at some personal cost over the years. In particular, it was a source of friction in his marriage to Jane, who professes the Christian faith. In her book, *Travelling to Infinity: My Life With Stephen*, she wrote:

> The damaging schism between religion and science seemed to have extended its reach into our very lives: Stephen would adamantly assert the blunt positivist stance which I found too depressing and too limiting to my view of the world, because I fervently needed to believe that there was more to life than the bald facts of the laws of physics and the day-to-day struggle for survival.

In a more acerbic moment, she even described her principal role within the marriage as being 'simply to tell him that he's not god'!

Hawking inevitably found himself on a collision course with established churches, most notably the Roman Catholic Church. He recalled, for instance, a conference hosted by the Vatican on the subject of cosmology during the papacy of John Paul II. In an article on Galileo in 2009, Hawking described how the Pope had given permission 'to study the beginning of the universe and where it began' but warned delegates that they 'should not inquire into the beginning itself, because that was the moment of Creation and the work of God'. 'I was glad he didn't realize I had presented a paper at the conference suggesting how the universe began,' Hawking said. 'I didn't fancy the thought of being handed over to the Inquisition, like Galileo.'

The question of whether there is a God or not will roll on for as long as there are humans around to debate it. Not even a mind as brilliant as Hawking's could claim to know for sure. While he clearly believed the scientific evidence argues heavily against the notion of a deity, there can be no empirical proof to say he was right. Faith, after all, is by definition an expression of belief, not fact. And while Hawking took us closer than anyone before to understanding how our universe came about, the question of why (if indeed there is a reason) remains as elusive as ever. In his 1999 work, *Alien Dawn*, the author and philosopher Colin Wilson summed it up like this:

I experience … [a] sense of absurdity when I listen to a cosmologist like Stephen Hawking telling us that the universe began with a big bang fifteen billion years ago, and that physics will shortly create a 'theory of everything' that will answer every possible question about our universe; this entails the corollary that God is an unnecessary hypothesis. Then I think of the day when I suddenly realized that I did not know where space ended, and it becomes obvious that Hawking is also burying his head in the sand. God may be an unnecessary hypothesis for all I know, and I do not have the least objection to Hawking dispensing with him, but until we can understand why there is existence rather than nonexistence, then we simply have no right to make such statements. It is unscientific.

In the early seventeenth century, Francis Bacon asserted in *Of the Advancement and Proficience of Learning, Divine and Human* that '… if any man think, by his enquiries after material things, to discover the nature or will of God, he is indeed spoiled by vain philosophy'. But it is perhaps best left to Jane Hawking, interviewed in the *Daily Telegraph* in 2015, to sum up the knotty problem at the heart of her ex-husband's take on religious belief:

Stephen made quite a point of keeping me guessing as to whether he was agnostic or atheist, but I liked to trip him up. I remember once asking him how he knew which theory to work on, to which he replied, 'Well you have to take a leap of faith in choosing the one that you think is going to be most productive.' I said, 'Really? I thought faith had no part to play in physics?'

# Hawking's Legacy

'Hawking acts like a great counter force against anti-intellectual movements. He takes complex scientific principles and explains them so the general public can understand and, more importantly, appreciate the science behind them.'

MICHAEL VENABLES, *ARS TECHNICA*, 2012

So how will Hawking be remembered? Will he be considered as right up there among the pantheon of greats or will he rather be seen as the celebrity scientist extraordinaire? The true nature of his legacy will likely remain uncertain for decades, or perhaps even centuries, to come. Only as his theories meet or fail empirical testing will his fate be decided, and our ability to test in this manner remains a long way off.

Nonetheless, we may say with some confidence that he was among the finest theoretical scientists of his era, even if it is more difficult to say whether he is the best of them all and the natural successor to the likes of Newton and Einstein. Certainly, there are some very credible advocates who claim he deserves elevation to the highest ranks. Prominent among them is Kip Thorne, who has said of Hawking that he rates 'besides Einstein, as the best in our field'.

The 'Establishment' most definitely recognized his extraordinary talents early on – a rather ironic turn of events given Hawking's anti-Establishment tendencies.

He was inducted into the Royal Society in 1974 for, in the institution's rather sober language, 'major contributions to the field of general relativity. These derive from a deep understanding of what is relevant to physics and astronomy, and especially from a mastery of wholly new mathematical techniques. Following the pioneering work of Penrose he established, partly alone and partly in collaboration with Penrose, a series of successively stronger theorems establishing the fundamental result that all realistic cosmological models must possess singularities. Using similar techniques, Hawking proved the basic theorems on the laws governing black holes: that stationary solutions of Einstein's equations with smooth event horizons must necessarily be axisymmetric; and that in the evolution and interaction of black holes, the total surface area of the event horizons must increase …'

Countless more official honours followed, including a CBE in the UK and the Medal of Freedom in the US. Meanwhile, handfuls of honorary degrees and the naming of several academic buildings in his honour attest to the high esteem in which his academic contemporaries held him. It also emerged that he turned down a knighthood from the British government in the 1990s, reputedly over his dissatisfaction with national science funding policies.

Yet there remain the naysayers, those who claim that the unprecedented success of *A Brief History of Time* created a phenomenon with a momentum of its own. His critics suggest that he was elevated from his rightful place in the swim with other leading academics working at the same time as him, instead being transformed into some sort of oracle of all things scientific. Take, for instance, the assertions of science writer Philip Ball, who penned an article for *Prospect Magazine* in 2010 under the headline 'The Hawking Delusion':

> Most people will be astonished to hear that Hawking is not rated by his peers among the top ten physicists even of the twentieth century, let alone of all time … Hawking is extremely smart, but so are others, and he is a long way from being Einstein's successor. More importantly, Hawking has no reputation among scientists as a deep thinker. There is nothing especially profound in what he has said to date about the social and philosophical implications of science in general and cosmology in particular.

Yet even those who argue that Hawking is not the 'genius' of folklore struggle to name anyone who has done more to further public engagement with the sciences. Indeed, Judy Bachrach, writing in *Vanity Fair* in 2004, claimed it is his 'gift for personalizing infinity that accounts for Hawking's success'. A year later Hawking spoke to the *Guardian* about the importance of communication. 'I think that it's important for scientists to explain their work, particularly in cosmo-logy,' he said. 'This now answers many questions once asked of religion.'

His innate ability to make even the most inaccessible science seem within the reach of the layman was evident from early on. Jane Hawking was one of the first to register it, as she told the *Daily Telegraph*:

> ... Stephen's intelligence fascinated me. I was no mathematician myself and hopeless at physics and yet he could explain things to me. We would look up at the night sky together, and although Stephen wasn't actually very good at detecting constellations, he would tell me about the expanding universe and the possibility of it contracting again and describe a star collapsing in on itself to form a black hole in a way that was quite easy to understand.

Prior to Hawking, Carl Sagan was arguably the most successful popularizer of science in the twentieth century. Yet Hawking came to surpass even his reach, and has also given impetus to a new generation of science communicators. Prominent among this new breed is Professor Brian Cox, a former minor pop star and the current go-to physicist for TV producers intent on making serious science programmes. He freely admits the debt he owes to Hawking, saying in 2012:'In terms of popularizing physics, I wouldn't be doing what I do now were it not for him.' So what does he believe was Hawking's secret? 'He ... realizes that it is not just our job to "do" physics but to explain it, as it is something which fascinates people.'

Even in death, Hawking is the most famous scientist in the world. While acknowledging that the jury remains out on the precise nature of his long-term contribution to the development of theoretical physics, only the truly churlish suggest that he has been anything other than a major player over several decades in a golden age for cosmology. Martin Rees, not generally prone to overstatement, pegged him in 2012 as 'one of the top ten living theoretical physicists'. Meanwhile, Kip Thorne told *Wired* magazine in 2015: 'He has pioneered completely new areas in physics. There were several key junctures in his career when he'd make a huge breakthrough and everybody else was struggling to catch up or struggling to understand.'

When he died on 14 March 2018, Hawking undoubtedly left behind a vacuum in the scientific firmament. Or, more properly in his case, a black hole. In an obituary published in *Nature* magazine, it fell to Martin Rees to sum up his legacy. 'Stephen's expectations when he was diagnosed [with ALS],' he wrote, 'dropped to zero; he said that everything that had happened since had been a bonus. And what a bonus – for physics, for the millions enlightened by his books and for the even larger number inspired by his achievement against all the odds.'

But perhaps the last word should be left to Hawking himself. How did he see his own contribution? 'It has been a glorious time to be alive and doing research in theoretical physics. I'm happy if I have added something to our understanding of the universe.'

# Selected Bibliography

Boslough, John, *Stephen Hawking's Universe*, William Morrow (1985)

Cox, Brian and Forshaw, Jeff, *The Quantum Universe: Everything That Can Happen Does Happen*, Penguin (2012)

Ferguson, Kitty, *Stephen Hawking: His Life and Work*, Bantam Books (2011)

Filkin, David, *Stephen Hawking's Universe*, BBC Books (1997)

Gribbin, John and White, Michael, *Stephen Hawking: A Life in Science*, Plume (1992)

Hawking, Jane, *Travelling to Infinity: My Life With Stephen*, Alma Books (2007)

Hawking, Stephen, *A Brief History of Time*, Bantam Books (1988)

Hawking, Stephen, *Black Holes and Baby Universes and Other Essays*, Bantam Books (1993)

Hawking, Stephen, *The Universe in a Nutshell*, Bantam Books (2001)

Hawking, Stephen, *On the Shoulders of Giants: The Great Works of Physics and Astronomy*, Running Press (2002)

Hawking, Stephen, *God Created the Integers: The Mathematical Breakthroughs That Changed History*, Running Press (2005)

Hawking, Stephen, *My Brief History*, Bantam Books (2013)

Hawking, Stephen, *Brief Answers to the Big Questions*, John Murray (2018)

Hawking, Stephen and Ellis, George, *The Large Scale Structure of Space-Time*, Cambridge University Press (1973)

Hawking, Stephen and Hawking, Lucy, *George's Secret Key to the Universe*, Corgi Childrens (2008)

# Selected Bibliography

Hawking, Stephen and Hawking, Lucy, *George's Cosmic Treasure Hunt,* Corgi Childrens (2010)

Hawking, Stephen and Hawking, Lucy, *George and the Big Bang*, Corgi Childrens (2012)

Hawking, Stephen and Hawking, Lucy, *George and the Unbreakable Code*, Corgi Childrens (2015)

Hawking, Stephen and Mlodinow, Leonard, *A Briefer History of Time*, Bantam Books (2005)

Hawking, Stephen and Mlodinow, Leonard, *The Grand Design*, Bantam Books (2010)

Hawking, Stephen and Penrose, Roger, *The Nature of Space and Time*, Princeton University Press (1996)

Krauss, Lawrence M., *The Physics of Star Trek*, HarperCollins (1996)

Susskind, Leonard, *The Black Hole War: My Battle With Stephen Hawking to Make the World Safe for Quantum Mechanics*, Back Bay Books (2009)

Thorpe, K. S., *Black Holes and Time Warps*, W. W. Norton & Co (1994)